Universitext

Springer
*New York
Berlin
Heidelberg
Barcelona
Budapest
Hong Kong
London
Milan
Paris
Santa Clara
Singapore
Tokyo*

D1153618

Universitext

Editors (North America): S. Axler, F.W. Gehring, and P.R. Halmos

Aksoy/Khamsi: Nonstandard Methods in Fixed Point Theory
Aupetit: A Primer on Spectral Theory
Booss/Bleecker: Topology and Analysis
Borkar: Probability Theory; An Advanced Course
Carleson/Gamelin: Complex Dynamics
Cecil: Lie Sphere Geometry: With Applications to Submanifolds
Chae: Lebesgue Integration (2nd ed.)
Charlap: Bieberbach Groups and Flat Manifolds
Chern: Complex Manifolds Without Potential Theory
Cohn: A Classical Invitation to Algebraic Numbers and Class Fields
Curtis: Abstract Linear Algebra
Curtis: Matrix Groups
DiBenedetto: Degenerate Parabolic Equations
Dimca: Singularities and Topology of Hypersurfaces
Edwards: A Formal Background to Mathematics I a/b
Edwards: A Formal Background to Mathematics II a/b
Foulds: Graph Theory Applications
Gardiner: A First Course in Group Theory
Gårding/Tambour: Algebra for Computer Science
Goldblatt: Orthogonality and Spacetime Geometry
Hahn: Quadratic Algebras, Clifford Algebras, and Arithmetic Witt Groups
Holmgren: A First Course in Discrete Dynamical Systems
Howe/Tan: Non-Abelian Harmonic Analysis: Applications of $SL(2, R)$
Howes: Modern Analysis and Topology
Humi/Miller: Second Course in Ordinary Differential Equations
Hurwitz/Kritikos: Lectures on Number Theory
Jennings: Modern Geometry with Applications
Jones/Morris/Pearson: Abstract Algebra and Famous Impossibilities
Kannan/Krueger: Advanced Real Analysis
Kelly/Matthews: The Non-Euclidean Hyperbolic Plane
Kostrikin: Introduction to Algebra
Luecking/Rubel: Complex Analysis: A Functional Analysis Approach
MacLane/Moerdijk: Sheaves in Geometry and Logic
Marcus: Number Fields
McCarthy: Introduction to Arithmetical Functions
Meyer: Essential Mathematics for Applied Fields
Mines/Richman/Ruitenburg: A Course in Constructive Algebra
Moise: Introductory Problems Course in Analysis and Topology
Morris: Introduction to Game Theory
Porter/Woods: Extensions and Absolutes of Hausdorff Spaces
Ramsay/Richtmyer: Introduction to Hyperbolic Geometry
Reisel: Elementary Theory of Metric Spaces
Rickart: Natural Function Algebras
Rotman: Galois Theory
Rubel/Colliander: Entire and Meromorphic Functions

(continued after index)

Lee A. Rubel

With assistance from James E. Colliander

Entire and Meromorphic Functions

Springer

Lee A. Rubel
Department of Mathematics
University of Illinois, Urbana–Champaign
Urbana, IL 61801-2917
USA
(deceased)

James E. Colliander
Department of Mathematics
University of Illinois, Urbana–Champaign
Urbana, IL 61801-2917
USA

Mathematics Subject Classification (1991): 30Dxx, 30D35

Library of Congress Cataloging-in-Publication Data
Rubel, Lee A.
    Entire and meromorphic functions / Lee A. Rubel with assistance
from James E. Colliander.
        p.   cm. — (Universitext)
    Includes bibliographical references and index.
    ISBN 0-387-94510-5 (softcover : alk. paper)
    1. Functions, Entire. 2. Functions, Meromorphic. 3. Nevanlinna
theory. I. Colliander, James E. II. Title.
    QA353.E5R83   1995
    515'.98—dc20   95-44887

Printed on acid-free paper.

With 2 illustrations.

Production managed by Laura Carlson; manufacturing supervised by Jacqui Ashri.
Camera-ready copy prepared using the authors' AMS-TeX files.
Printed and bound by R.R. Donnelley & Sons, Harrisonburg, VA
Printed in the United States of America.

9 8 7 6 5 4 3 2 1

ISBN 0-387-94510-5 Springer-Verlag New York Berlin Heidelberg   SPIN 10424824

Dedicated to the Memory of Steven B. Bank
Student, Colleague, Teacher, Friend

Contents

1
Introduction

Mathematics is a beautiful subject, and entire functions is its most beautiful branch. Every aspect of mathematics enters into it, from analysis, algebra, and geometry all the way to differential equations and logic.

For example, my favorite theorem in all of mathematics is a theorem of R. Nevanlinna that two functions, meromorphic in the whole complex plane, that share five values must be identical. For real functions, there is nothing that even remotely corresponds to this.

This book is an introduction to the theory of entire and meromorphic functions, with a heavy emphasis on Nevanlinna theory, otherwise known as *value-distribution theory*. Things included here that occur in no other book (that we are aware of) are the Fourier series method for entire and meromorphic functions, a study of integer valued entire functions, the Malliavin-Rubel extension of Carlson's Theorem (the "sampling theorem"), and the first-order theory of the ring of all entire functions, and a final chapter on Tarski's "High School Algebra Problem," a topic from mathematical logic that connects with entire functions.

This book grew out of a set of classroom notes for a course given at the University of Illinois in 1963, but they have been much changed, corrected, expanded, and updated, partially for a similar course at the same place in 1993. My thanks to the many students who prepared notes and have given corrections and comments.

In order to discover and prove interesting, deep, or powerful theorems in this area, what we most need is more *examples* of interesting meromorphic functions—I would guess that the number of fundamentally different examples known is about 20 or 30. One promising source of such examples is the Painlevé transcendents (see [14], pp. 438–444, and [29]).

However, in spite of a growing literature on these functions, the unfortunate fact is that the "proofs" are incomplete and not rigorous—indeed, there still is not a satisfactory proof that the Painlevé transcendents of even the first kind (i.e., solutions of $w'' = 6w^2 + z$) are meromorphic in the full complex plane. Basic notions like "fixed singularity" and "movable singularity," however intuitively appealing, have never been given rigorous definitions.

It is hard to see how this lamentable situation will improve since, the world being as it is, there is little "glory" attached to proving theorems that have already been "proved."

The subject of entire and meromorphic functions has been growing for many decades, and will continue to grow forever. It is hoped that this book will give the novice reader a good introduction to the subject, or the expert some new insights. This book could "easily" have been four or five times it length, since the subject is so extensive, but to use my favorite saying, "enough is too much."

LEE A. RUBEL

Lee Rubel died on March 25, 1995. As my teacher, the way his personality merged into his mathematics always inspired me. I sincerely hope that readers of this book find similar inspiration.

JAMES E. COLLIANDER

October 18, 1995

2
The Riemann-Stieltjes Integral

We give here a brief summary of some of the basic facts about the Riemann-Stieltjes integral. Those unfamiliar with the subject are urged to read Chapter 9 of *Mathematical Analysis* by Apostol [1].

Throughout this section, a and b are real numbers, usually $a < b$, and f and α are real-valued functions defined on the closed interval $[a, b]$. When f and α are suitably restricted, we will define $\int_a^b f \, d\alpha$ as the Riemann-Stieltjes integral of f with respect to α. When $\alpha(x) = x$ for all x in $[a, b]$, $\int_a^b f \, d\alpha$ is the ordinary Riemann integral of f, and many of the familiar properties of the Riemann integral extend to the Riemann-Stieltjes integral.

Definition. A *partition* of $[a, b]$ is an ordered $(n + 1)$-tuple

$$P = \{x_0, x_1, \ldots, x_n\} \quad \text{with} \quad x_0 = a,$$

$x_n = b$, and $x_{j-1} < x_j$ for $j = 1, \ldots, n$.

Definition. A *selection* σ from a partition P is an ordered n-tuple $\sigma = \{t_1, \ldots, t_n\}$ such that

$$x_{j-1} \le t_j \le x_j \quad \text{for} \quad j = 1, \ldots, n.$$

Definition. A *Riemann-Stieltjes sum* (of f with respect to α) is a sum of the form

$$S(P, \sigma : f, \alpha) = \sum_{k=1}^{n} f(t_k)\{\alpha(x_k) - \alpha(x_{k-1})\}.$$

Definition. P' is a refinement of P; written $P \subset P'$, means that each x that occurs in P also occurs in P'.

Definition. f is *integrable with respect to* α; written $f \in R(\alpha)$, means that there exists a number A such that, for each $\epsilon > 0$, there exists a partition P_ϵ such that $P_\epsilon \subset P$; and if σ is a selection from P, then

$$|S(P, \sigma : f, \alpha) - A| < \epsilon.$$

It is easily seen that A is uniquely determined if $f \in R(\alpha)$, so that we may write

$$\int_a^b f \, d\alpha = \int_a^b f(x) \, d\alpha(x) = A.$$

Theorem. *If f is continuous on $[a, b]$ and α is monotone on $[a, b]$, then $f \in R(\alpha)$ and $\alpha \in R(f)$.*

Theorem (Integration by parts). *If $f \in R(\alpha)$, then $\alpha \in R(f)$ and*

$$\int_a^b f \, d\alpha = f(b)\alpha(b) - f(a)\alpha(a) - \int_a^b \alpha \, df.$$

Under suitable hypotheses, the usual linearity properties and formulas for "change of variables" hold, e.g.,

$$\int_a^b (Af + Bg) d\alpha = A \int_a^b f d\alpha + B \int_a^b g d\alpha$$

$$\int_a^b f d(A\alpha + B\beta) = A \int_a^b f d\alpha + B \int_a^b f d\beta$$

$$\int_a^c = \int_a^b + \int_b^c$$

$$\int_a^b f(x) d\alpha(x) = \int_a^b f(x)\alpha'(x) dx$$

$$\int_a^b f d\alpha = \int_c^d h d\beta$$

if

$$h = f \circ g, \quad \beta = \alpha \circ g,$$
$$a = g(c), \quad b = g(d),$$

where $(f \circ g)(x) = f(g(x))$.

Theorem. *If α is a step function that jumps α_k at points x_k, $k = 1$, $2, \ldots, n$ in $[a, b]$ and f is continuous on $[a, b]$, then*

$$\int_a^b f(x) d\alpha(x) = \sum_{k=1}^n f(x_k) \alpha_k.$$

Remark. Given a finite sum $\sum a_k b_k$, we may write

$$\sum a_k b_k = \int a \, dB,$$

where B is a step function that jumps b_k at x_k and a is a continuous function such that $a(x_k) = a_k$. Integration by parts now becomes "partial summation" and we have the formula

$$a_0 b_0 + a_1 b_1 + \cdots a_N b_N = A_0(b_0 - b_1) + \cdots + A_{N-1}(b_{N-1} - b_N) + A_N b_N,$$

where $A_k = a_0 + a_1 + \cdots + a_k$.

First Mean Value Theorem. *If α is nondecreasing on $[a, b]$, $f \in R(\alpha)$, and*

$$M = \sup\{f(x) : x \in [a, b]\}$$
$$m = \inf\{f(x) : x \in [a, b]\},$$

then there is a number c, $m \le c \le M$, such that

$$\int_a^b f(x) d\alpha(x) = c \int_a^b d\alpha(x) = c[\alpha(b) - \alpha(a)].$$

If f is continuous, then $c = f(x_0)$ for some $x_0 \in [a, b]$. In particular,

$$m \int_a^b d\alpha(x) \le \int_a^b f(x) d\alpha(x) \le M \int_a^b d\alpha(x).$$

Second Mean Value Theorem. *Suppose that α is continuous and f is nondecreasing on $[a, b]$. Then there is an $x_0 \in [a, b]$ such that*

$$\int_a^b f(x) d\alpha(x) = f(a) \int_a^{x_0} d\alpha(x) + f(b) \int_{x_0}^b d\alpha(x).$$

Corollary (Bonnet's theorem). *If g is continuous and f is nonnegative and nondecreasing on $[a, b]$, then*

$$\int_a^b f(x) g(x) dx = f(b) \int_{x_0}^b g(x) dx$$

for a suitable choice of $x_0 \in [a, b]$.

3
Jensen's Theorem and Applications

One of our most useful tools is Jensen's Theorem, which can be used to relate the distribution of zeros of an entire function to its growth. We prove Jensen's Theorem using the Gauss Mean Value Theorem.

Gauss Mean Value Theorem. *Suppose u is a harmonic function in $\mathbb{D}$. Then the value of u at the center is equal to the average of the boundary values of u. That is,*

$$(3.1) \qquad u(0) = \frac{1}{2\pi} \int_0^{2\pi} u(e^{it}) \, dt$$

Proof. Form the analytic function $f(z)$ whose real part is $u(x, y)$. Apply Cauchy's integral formula to evaluate f at zero, then take the real part and the theorem is proved.

Jensen's Theorem. *If f is meromorphic in $|z| \leq R$, if $r < R$, and if*

$$f(z) = a_k z^k + a_{k+1} z^{k+1} + \ldots \quad (a_k \neq 0)$$

is the Laurent expansion of f around zero, then

$$(3.2) \quad \frac{1}{2\pi} \int_{-\pi}^{\pi} \log |f(re^{i\theta})| \, d\theta = \log |a_k| + \sum_{r_n \leq r} \log \frac{r}{r_n} - \sum_{\rho_n \leq r} \log \frac{r}{\rho_n} + k \log r,$$

where the zeros of f are $z_j = r_j e^{i\theta_j}$ and the poles of f are $w_j = \rho_j e^{i\phi_j}$, not counting zeros or poles at the origin.

Proof. With no loss in generality, assume $f(0) = 1$ and $R = 1$:

$$(3.3) \qquad F(z) = \frac{f(z)}{\prod_n \left(\dfrac{z - z_n}{1 - \bar{z}_n z} \right)} \cdot \prod_n \left(\frac{z - w_n}{1 - \bar{w}_n z} \right).$$

F is an analytic function with no zeros and no poles. Since $\log |F|$ is a harmonic function in the disk, the Gauss Mean Value (3.1) implies

$$(3.4) \qquad \frac{1}{2\pi} \int_{-\pi}^{\pi} \log |F(re^{i\theta})| \, d\theta = \log |f(0)| + \sum \log \rho_n - \sum \log r_n.$$

But $\frac{1}{2\pi} \int_{-\pi}^{\pi} \log |F(re^{i\theta})| \, d\theta = \frac{1}{2\pi} \int_{-\pi}^{\pi} \log |f(re^{i\theta})| $ since $\left| \frac{z - z_0}{1 - \bar{z}_0 z} \right| = 1$ on $|z| = 1$ if $|z_0| < 1$.

Let $n(r, f)$ be the number of poles of f in the closed disk $|z| \le r$, counted according to multiplicity. Thus $n\left(r, \frac{1}{f-a}\right)$ counts the number of a-points [an a-point is a point z satisfying $f(z) = a$] of f. Let $n(r) = n\left(r, \frac{1}{f}\right)$, which counts the zeros of f. Let $k^+ = \max(k, 0)$ and $k^- = -\min(k, 0)$, so that $k^+ - k^- = k$. Then,

$$(3.5) \qquad \sum_{r_n \le r} \log \frac{r}{r_n} + k^+ \log r = \int_{0+}^{r} \log \frac{r}{t} d\{n(t) - n(0)\} + k^+ \log r$$

with $u = \log \frac{r}{t}$, $du = -\frac{dt}{t}$, and $v = n(t) - n(0)$. An integration by parts gives[†]

$$[n(t) - n(0)] \log \frac{r}{t} \Big|_{0+}^{r} + \int_{0+}^{r} \frac{n(t) - n(0)}{t} \, dt + k^+ \log r$$

$$= \int_{0+}^{r} \frac{n(t) - n(0)}{t} \, dt + k^+ \log r.$$

We define:

$$N\left(r, \frac{1}{f}\right) \equiv k^+ \log r + \int_{0+}^{r} \frac{n\left(t, \frac{1}{f}\right) - n\left(0, \frac{1}{f}\right)}{t} \, dt = \sum_{r_n \le r} \log \frac{r}{r_n} + k^+ \log r$$

$$N(r, f) \equiv k^- \log r + \int_{0+}^{r} \frac{n(t, f) - n(0, t)}{t} \, dt = \sum_{\rho_n \le r} \log \frac{r}{\rho_n} + k^- \log r.$$

Remarks. $n(r, f)$ counts the number of poles of f in the disk $|z| \le r$. $N(r, f)$ is a useful average of the counting function. We usually normalize f so that $f(0) = 1$, and in this case $k^+ = k^- = n(0, f) = n\left(0, \frac{1}{f}\right) = \log |a_k| = 0$.

As a typical application of Jensen's Theorem, we prove the next result.

[†]The 0^+ indicates that we integrate from, say, ϵ to r, where ϵ is smaller than the smallest positive modulus of a zero.

Theorem. *Given $z_1, z_2, \ldots$ with $0 < |z_j| = r_j < 1$, there exists a bounded holomorphic function f in the unit disk whose zeros are precisely the z_j if and only if*

$$\sum (1 - r_j) < \infty.$$

Proof. If such an f exists, we may suppose $f(0) = 1$ and apply Jensen's Theorem:

$$N\left(r, \frac{1}{f}\right) = \sum_{r_j \leq r} \log \frac{r}{r_j} = \frac{1}{2\pi} \int_{-\pi}^{\pi} \log |f(re^{i\theta})| \, d\theta.$$

Since f is bounded, $N\left(r, \frac{1}{f}\right)$ is bounded. $N\left(r, \frac{1}{f}\right)$ is nondecreasing, so $\sum \log\left(\frac{1}{r_j}\right) < \infty$. But (exercise) $\sum \log\left(\frac{1}{r_j}\right) < \infty$ if and only if $\sum \log(1 - r_j) < \infty$.

In the other direction, suppose $\sum \log\left(\frac{1}{r_j}\right) < \infty$ and let

$$P_n(z) = \prod_{j=1}^{n} \frac{z - z_j}{1 - \bar{z}_j z}.$$

The P_n form a normal family since $|P_n(z)| \leq 1$ for all n and all $z \in \mathbb{D}$. Passing to a subsequence if necessary, the P_n converge uniformly on compact subsets of $\mathbb{D}$ to a bounded holomorphic function f. Can f be identically zero? Since $|P_n(0)| = \left|\prod_1^n(-z_j)\right| = \prod_1^n r_j$, we have $|f(0)| \geq \prod_1^\infty r_j$. Hence $|f(0)| > 0$ since $\log \prod_1^\infty \left(\frac{1}{r_j}\right) = \sum_1^\infty \log \frac{1}{r_j} < \infty$.

4
The First Fundamental Theorem of Nevanlinna Theory

Rewriting Jensen's Theorem, we get

$$(4.1) \qquad \frac{1}{2\pi} \int_{-\pi}^{\pi} \log |f(re^{i\theta})| \, d\theta = \log |a_k| + N\left(r, \frac{1}{f}\right) - N(r, f),$$

where N is a kind of average number of poles of f.

For positive numbers x, let us write

$$\log^+ x = \max(0, \log x) = \log[\max(1, x)]$$

so that

$$\log x = \log^+ x - \log^+ \frac{1}{x}.$$

We now list some simple properties of $\log^+$:

(a) $\log^+(x_1 \cdot x_2 \cdot \ldots \cdot x_n) \leq \log^+ x_1 + \log^+ x_2 + \cdots + \log^+ x_n$.

(b) $\log^+(x_1 + x_2 + \cdots + x_n) \leq \log^+ x_1 + \log^+ x_2 + \cdots + \log^+ x_n + \log n$.

In particular,

(c) $\log^+(x_1 + x_2) \leq \log^+ x_1 + \log^+ x_2 + \log 2$.

From (c), we get

$$\log^+ |x - a| \leq \log^+ |x| + \log^+ |a| + \log 2$$
$$\log^+ |x| \leq \log^+ |x - a| + \log^+ |a| + \log 2,$$

so that

(d) $\left| \log^+ |x - a| - \log^+ |x| \right| \leq \log^+ |a| + \log 2$.

We may write

$$\frac{1}{2\pi}\int_{-\pi}^{\pi}\log|f(re^{i\theta})|\,d\theta = \frac{1}{2\pi}\int_{-\pi}^{\pi}\log^+|f(re^{i\theta})|\,d\theta$$

(4.2)
$$-\frac{1}{2\pi}\int_{-\pi}^{\pi}\log^+\left|\frac{1}{f(re^{i\theta})}\right|\,d\theta.$$

Let $m(r,f) = \frac{1}{2\pi}\int_{-\pi}^{\pi}\log^+|f(re^{i\theta})|\,d\theta$. Then we may rewrite Jensen's Theorem as

$$m(r,f) - m\left(r,\frac{1}{f}\right) = \log|a_k| + N\left(r,\frac{1}{f}\right) - N(r,f)$$

or

(4.3) $$m(r,f) + N(r,f) = \log|a_k| + m\left(r,\frac{1}{f}\right) + N\left(r,\frac{1}{f}\right).$$

Notice that $N(r,f)$ counts the poles of f (with a certain kind of averaging) that is the averaged number of times f takes the value ∞, while $m(r,f)$ measures the tendency of f to take the value ∞. Hence, the quantity $m(r,f) + N(r,f)$ measures, in some sense, the total affinity of f for the value ∞. Similarly, $m\left(r,\frac{1}{f}\right) + N\left(r,\frac{1}{f}\right)$ measures the total affinity of f for the value zero. So the above version of Jensen's Theorem asserts that the total affinity of f for ∞ is the same as the total affinity of f for the value zero, modulo a bounded function of r. The first fundamental theorem is based on the observation that, for any constant a, the affinity of $f - a$ for ∞ is essentially the same as that for f, while the affinity of $f - a$ for zero is, of course, the affinity of f for a. The theorem states that $m\left(r,\frac{1}{f-a}\right) + N\left(r,\frac{1}{f-a}\right)$ is independent of a, modulo a bounded function of r. Here we use the convention that if $a = \infty$, then $\frac{1}{f-a}$ means f.

Fix $a \in \mathbb{C}$. Then $N(r,f) = N(r,f-a)$ since z is a pole of f if and only if z is a pole of $f - a$. From property (d) of $\log^+$ we obtain

$$|m(r,f-a) - m(r,f)| \le \log^+|a| + \log 2.$$

We define

$$T(r,f) = m(r,f) + N(r,f).$$

T is called the *(Nevanlinna) characteristic* of f.

From Jensen's Theorem, we have

$$T(r,f) = T\left(r,\frac{1}{f}\right) + \phi(r)$$

$$T(r,f-a) = T\left(r,\frac{1}{f-a}\right) + \phi(r;a),$$

where $\phi(r) = \phi(r;0) = \log|a_k|$ and $\phi(r;a) = \log|a_k(a)|$. Here $a_k(a)$ is the first nonvanishing coefficient in the Laurent expansion of $f - a$ at the origin. For each $a \in \mathbb{C}$, $\phi(r;a)$ is a bounded function of r.

First Fundamental Theorem of Nevanlinna Theory. *If f is meromorphic in $|z| < R$, where $0 < R \leq \infty$, then*

$$(4.4) \qquad T\left(r, \frac{1}{f-a}\right) = T(r,f) + \phi(r;a),$$

where $|\phi(r;a)| \leq \log^+|a| + \left|\log^+|a_k(z)|\right| + \log 2$ for all r with $0 \leq r < R$.

Proof. This is simply a rephrasing of Jensen's Theorem using the new notation we have introduced.

Since it is customary to work modulo bounded functions of r, we may sometimes abuse the notation and write things like $T(r,f) = T(r, f-a)$ when we mean only $T(r,f) = T(r, f-a) + O(1)$. The characteristic plays a central role in the theory of meromorphic (and entire) functions.

5
Elementary Properties of $T(r, f)$

In this chapter, we present basic properties of the characteristic function.

(5.1) $$T(r, h_1 h_2) \leq T(r, h_1) + T(r, h_2)$$

(5.2) $$T(r, h_1 + h_2) \leq T(r, h_1) + T(r, h_2) + \log 2$$

(5.3) $$T(r, \frac{h_1}{h_2}) \leq T(r, h_1) + T(r, h_2) + O(1)$$

(5.4) $$T(r, \pm f^n) = nT(r, f); \quad n = 1, 2, \ldots .$$

The proofs of (5.1), (5.2), and (5.4) are simple consequences of the properties of $\log^+$ and of the fact that $N(r, f)$ "counts poles," while (5.3) follows from (5.1) and the first fundamental theorem, which implies that

$$T(r, f) = T\left(r, \frac{1}{f}\right) + O(1)$$

(5.5)
$$T\left(r, \sum_0^n h_j h^j\right) = nT(r, h) + \sum_0^n T(r, h_j) + n \log 2.$$

The proof of (5.5) is by induction based on the identity:

$$\sum_0^n h_j h^j = h_0 + h \sum_1^n h_j h^{j-1}$$

to which (5.1) and (5.2) are applied in the obvious way.

Definition. To each function $\lambda(r)$ that is positive, continuous, and non-decreasing on $0 \leq r < \infty$, we associate the class Λ of functions f that are meromorphic in $|z| < \infty$ and that satisfy

$$T(r, f) \leq A\lambda(Br)$$

for positive constants A and B as $r \to R$. If f is entire and satisfies this condition, we say that f is an *entire function of finite λ-type*.

It is easy to verify that Λ is a field, and we call any such field a Λ-*field*.

Remarks. Some Λ-fields have been studied heavily. The case $\lambda(r) = \max(1, r^\rho)$ is especially important and we denote the corresponding Λ-field by Λ_ρ. In case $f \in \Lambda_\rho$, we say that f is of order at most ρ and of exponential type. In case $\rho = 1$, we say simply that f is of exponential type. The intersection, $\Omega_\rho = \bigcap_{\rho' > \rho} \Lambda_{\rho'}$, consists of all functions of order at most ρ. We shall discuss order and type in much greater detail later on. Notice that the next theorem implies that the fields of all meromorphic functions (i) of order at most ρ exponential type and (ii) of order at most ρ are algebraically closed in the field of all meromorphic functions on $\mathbb{C}$.

Theorem. *Each Λ-field is algebraically closed in the field of all meromorphic functions on $\mathbb{C}$.*

By this we mean the following. Suppose that f, f_0, $f_1, \ldots, f_n$ are meromorphic in $\mathbb{C}$, that f_n is not identically zero, that $f_j \in \Lambda$ for $j = 0, \ldots, n$, and that

(5.6) $$f_0 + f_1 f + f_2 f^2 + \cdots + f_n f^n = 0.$$

Then the theorem asserts that $f \in \Lambda$. Notice that we do not prove the existence of an f that satisfies (5.6).

Proof. From (5.6) we may write

$$T(r, f^n) = T\left(r, -\frac{1}{f_n} \sum_{j=0}^{n-1} f_j f^j\right).$$

Then by (5.3), (5.4), and (5.5) we get

$$T(r, f) \leq \sum_{j=0}^{n} T(r, f_j) + O(1) \quad \text{as} \quad r \to \infty$$

and thus $T(r, f) \leq A\lambda(Br)$ of appropriate constants A and B.

Definition. Let ℓ denote the ring of all of those functions that belong to Λ and are holomorphic in $\mathbb{C}$.

Theorem. *Each ℓ-ring is algebraically closed in the ring of all functions holomorphic in D_R.*

This follows directly from the preceding theorem.

We give a detailed proof of the theorem only outlined in [12, p. 54], reversing the notation for f and g.

Clunie's Theorem. *Let $f(z)$ be a transcendental entire function, let $g(z)$ be a nonconstant entire function, and let $\varphi(z) = f(g(z))$. Then*

$$\frac{T(r, \varphi)}{T(r, g)} \longrightarrow \infty$$

as $r \to \infty$.

Proof. We may and do assume that $f(w)$ has infinitely many distinct zeros at $w_1, w_2, \cdots \to \infty$. (Otherwise, we could replace f by $f - \lambda$ for a suitable constant λ. We are implicitly using the fact that if f is an entire function that takes each complex number a as a value only finitely many times, then f must be a polynomial. Take this as an exercise. [Hint: Hurwitz's Theorem, the Casorati-Weierstrass Theorem, and Liouville's Theorem.] This fact is also a consequence of several of the later results in this book, like Picard's theorem.) Then, for any integer P,

$$(5.7) \qquad N\left(r, \frac{1}{\varphi}\right) \geq \sum_{\nu=1}^{P} N\left(r, \frac{1}{g(w) - w_\nu}\right)$$

because the averaged counting function N is a monotone increasing function of the pole set. We also want

$$(5.8) \qquad m\left(r, \frac{1}{\varphi}\right) \geq \sum_{\nu=1}^{P} m\left(r, \frac{1}{g(w) - w_\nu}\right) - O(1).$$

Fix P and let

$$(5.9) \qquad 0 < \delta < \frac{1}{10} \min\{|w_i - w_j| : i \neq j, \quad i, j = 1, \ldots, P\}.$$

Write

$$(5.10) \qquad f(w) = (w - w_1)^{m_1} \cdots (w - w_p)^{m_p} \Phi(w),$$

where $\Phi(w)$ is nonzero at each w_i, and where we also choose δ so small that $\Phi(w) \neq 0$ for $0 < |w - w_i| \leq \delta$ for all $i - 1, 2, \ldots, P$. Say $|\Phi(w)| \geq \epsilon > 0$ for all w within δ of w_i, $i = 1, 2, \ldots, P$.

Define

$$(5.11) \qquad E = \bigcup_{i=1}^{n} \{z : |g(z) - w_i| \leq \delta\}.$$

For $z \in E$, we have

(5.12) $$\log^+ \frac{1}{|f(g(z))|} \geq \sum_{\nu=1}^{P} \log^+ |\frac{1}{g(z) - w_\nu}| - M$$

for a suitable constant M depending on P, δ, and ϵ. But

$$\frac{1}{2\pi} \int_{-\pi}^{\pi} \log^+ |\frac{1}{g(re^{i\theta}) - w_i}| \, d\theta$$

is asymptotic as $r \to \infty$, to

$$\frac{1}{2\pi} \int_{E_i} \log^+ |\frac{1}{g(re^{i\theta})}| \, d\theta,$$

where

$$E_i = \{\theta : |g(re^{i\theta}) - w_i| < \delta\}$$

because the integral over the remaining part is less than $\log^+ \frac{1}{\delta}$.

We conclude, using (5.12), that

$$T\left(r, \frac{1}{\varphi}\right) \geq PT(r, g) + O(1)$$

for any integer P, and the result follows.

6
The Cartan Formulation of the Characteristic

We begin with some remarks on convex functions.

Lemma. *If β is a nondecreasing function, then $B(t) = \int_a^t \beta(s)\, ds$ is a convex function of t.*

Corollary. *If α is a nondecreasing function, then $A(r) = \int_0^r \alpha(x)\frac{dx}{x}$ is a convex function of $\log r$, that is, $A(e^t)$ is a convex function of t.*

The corollary follows directly from the lemma since

$$A(e^t) = \int_0^{e^t} \alpha(x)\frac{dx}{x} = \int_{-\infty}^t \beta(s)\, ds, \quad \text{where} \quad \beta(s) = \alpha(e^s).$$

Of course, we assume that α is small enough near 0 so that the integral exists: In most of our applications, we will have $\alpha(x) = 0$ for $0 \le x \le x_0$ for some $x_0 > 0$. Formally, a condition that B is convex is that $B'(t)$ shall be nondecreasing, and here $B'(t) = \beta(t)$. Similarly, $rA'(r) = \alpha(r)$.

Proof. After a simple normalization, it is seen that we must prove that

$$\int_0^x \beta(s)\, ds \le x \int_0^1 \beta(s)\, ds \quad \text{for} \quad 0 \le x \le 1.$$

This will be the case if $\frac{1}{x}\int_0^x \beta(s)\, ds$ is nondecreasing. But $\frac{1}{x}\int_0^x \beta(s)\, ds = \int_0^1 \beta(xt)\, dt$, which is obviously nondecreasing since β is. Suppose $y > x$. Then we see

$$\int_0^1 \beta(yt)\, dt - \int_0^1 \beta(xt)\, dt = \int_0^1 [\beta(yt) - \beta(xt)]\, dt \ge 0.$$

Roughly speaking, B is a convex function if and only if B can be represented as in the lemma. Similarly, A is logarithmically convex if and only if A can be represented as in the corollary.

We now reformulate the first fundamental theorem following a procedure due to Cartan.

Theorem. *For a certain constant C,*

$$T(r, f) = C + \frac{1}{2\pi} \int_{-\pi}^{\pi} N\left(r, \frac{1}{f - e^{i\varphi}}\right) d\varphi$$

$$= C + \int_{0}^{r} \left\{ \frac{1}{2\pi} \int_{-\pi}^{\pi} n\left(t, \frac{1}{f - e^{i\varphi}}\right) d\varphi \right\} \frac{dt}{t}.$$

In particular, $T(r, f)$ is a nondecreasing convex function of $\log r$.

Proof. Apply Jensen's Theorem to the function $f - e^{i\varphi}$ for some real constant φ:

(6.1)
$$\frac{1}{2\pi} \int_{-\pi}^{\pi} \log\left|f(e^{i\theta}) - e^{i\varphi}\right| d\theta = \log|a_k(\varphi)| + N\left(r, \frac{1}{f - e^{i\varphi}}\right) - N(r, f).$$

Thinking of the Laurent series for $f - e^{i\varphi}$ and the definitions of N and $a_k(\varphi)$, it is obvious that
  (i) If $k < 0$, $a_k(\varphi) = a_k$ and $k(\varphi) = k$ for all $\varphi \in \mathbb{R}$.
  (ii) If $k = 0$, $a_k(\varphi) = a_k - e^{i\varphi}$ and $k(\varphi) = k = 0$ for all $\varphi \in \mathbb{R}$.
  (iii) If $k > 0$, $a_k(\varphi) = -e^{i\varphi}$ and $k(\varphi) = 0$. Therefore, since $k^+(\varphi) = 0$ for $\varphi \in \mathbb{R}$,

$$\frac{1}{2\pi} \int_{-\pi}^{\pi} k^+(\varphi) \, d\varphi = 0.$$

With at most one exception [namely if $f(0) = e^{i\varphi}$ for some $\varphi \in \mathbb{R}$] we have

$$n\left(0, \frac{1}{f - e^{i\varphi}}\right) = 0.$$

Now, by Jensen's Theorem for any constant b, we get

(6.2)
$$\frac{1}{2\pi} \int_{-\pi}^{\pi} \log|b - e^{i\varphi}| \, d\pi = \log^+ |b|.$$

(Check the cases where $|b| \geq 1$ and $|b| < 1$.)

Let us integrate (6.1) with respect to φ:

(6.3)
$$\frac{1}{2\pi} \int_{-\pi}^{\pi} \left\{ \frac{1}{2\pi} \int_{-\pi}^{\pi} \log|f(re^{i\theta}) - e^{i\varphi}| \, d\theta \right\} d\varphi$$

$$= \frac{1}{2\pi} \int_{-\pi}^{\pi} \log|a_k(\varphi)| \, d\varphi + \frac{1}{2\pi} \int_{-\pi}^{\pi} N\left(r, \frac{1}{f - e^{i\varphi}}\right) d\varphi - N(r, f).$$

Applying Fubini's Theorem to the left-hand side (LHS) and using (6.2) yields

$$\frac{1}{2\pi}\int_{-\pi}^{\pi}\frac{1}{2\pi}\int_{-\pi}^{\pi}\log|f(re^{i\theta})-e^{i\varphi}|\ d\theta\ d\varphi = \frac{1}{2\pi}\int_{-\pi}^{\pi}m(r,f)\ d\theta = m(r,f).$$

Hence we obtain the result:
(6.4)
$$m(r,f)+N(r,f) = \frac{1}{2\pi}\int_{-\pi}^{\pi}\log|a_k(\varphi)|\ d\varphi + \frac{1}{2\pi}\int_{-\pi}^{\pi}N\left(r,\frac{1}{f-e^{i\varphi}}\right)\ d\varphi.$$

The first integral is a constant and we are done.

Remark. We could define $T^0(r) = \frac{1}{2\pi}\int_{-\pi}^{\pi}N\left(r,\frac{1}{f-e^{i\varphi}}\right)\ d\varphi$ as the Cartan characteristic of f, but by a minor abuse of notation we shall write $T(r) = \frac{1}{2\pi}\int_{-\pi}^{\pi}N\left(r,\frac{1}{f-e^{i\varphi}}\right)\ d\varphi$ since we customarily work modulo bounded functions of r anyway.

Interpretation. Roughly, we have

$$r\frac{dT}{dr} = L(r),$$

where

$$L(r) = \frac{1}{2\pi}\int_{-\pi}^{\pi}n\left(r,\frac{1}{f-e^{i\varphi}}\right)\ d\varphi.$$

What does the function L measure? The function $f(re^{i\theta})$ may be considered as a mapping of the circumference $\partial\mathbb{D}_r = \{z : |z| = r\}$ into the Riemann sphere. $n\left(r,\frac{1}{f-e^{i\varphi}}\right)$ counts the number of times the point $e^{i\varphi}$ is covered by this map. Hence, $\int_{-\pi}^{\pi}n\left(r,\frac{1}{f-e^{i\varphi}}\right)\ d\varphi$ measures the total arc length of the unit circumference (counting multiplicity) covered by the mapping f. In other words, the more heavily the mapping f covers the unit circumference, the faster T grows. Thus, T measures the covering properties of f.

We outline here another characteristic, the *Ahlfors-Shimizu characteristic* $T_A(r,f)$, which behaves much like the Nevanlinna characteristic but which has as an enlightening geometric interpretation. For full details, see, from which our presentation is abstracted.

We define

$$T_A(r,a) = M_A(r,a) + N(r,a),$$

where

$$N(r,a) = \int_0^r\frac{n(t,a)}{t}\ dt,$$

as before, but

$$M_A(r, a) = \frac{1}{2\pi} \int_0^{2\pi} \log \frac{1}{[w, a]} \, d\theta,$$

where

$$[w, a] = \frac{|w - a|}{\sqrt{1 + |a|^2}\sqrt{1 + |w|^2}} \quad \text{and} \quad [\infty, a] = \frac{1}{\sqrt{1 + |a|^2}}.$$

Now $[w, a]$ is the distance on the Riemann sphere between the points on the sphere to which w and a correspond via stereographic projection.

It is easy to see that

$$\left| T(r, \infty) - T_A(r, \infty) - \log^+ |f(a)| \right| \le \frac{1}{2} \log 2.$$

By Green's Theorem, one can show that

$$\frac{r}{2\pi} \frac{d}{dr} \int_0^{2\pi} \log \sqrt{1 + |f(re^{i\theta})|^2} \, d\theta + n(r, f) = \frac{1}{\pi} \int_0^r \int_0^{2\pi} \frac{|f'(pe^{i\theta})|^2 p \, dp \, d\theta}{[1 + |f(pe^{i\theta})|^2]^2}.$$

Denote the right side by $A(r)$, divide by r, and integrate the resulting identity from 0 to r to get

$$\int_0^r \frac{A(t)}{t} \, dt = N(r, f) + \frac{1}{2\pi} \int_0^{2\pi} \log \sqrt{1 + |f(re^{i\theta})|^2} \, d\theta - \log \sqrt{1 + |f(0)|^2}.$$

If we make a rotation of the Riemann sphere, which corresponds to the transformation

$$\bar{w} = \frac{1 + \bar{a}w}{w - a},$$

where $w = f(z)$, and call the resulting function $\bar{w} = F(z)$, we can derive the first fundamental theorem for the Ahlfors-Shimizu characteristic.

Theorem. *If f is meromorphic in $|z| < R$, where $0 < R \le \infty$, then for every finite or infinite a and r with $0 < r \le R$ we have,*

$$T_A(r, \infty) = \int_0^r \frac{A(t)}{t} \, dt = N(r, a) + M_A(r, a) - M_A(0, a).$$

For the geometrical interpretation of T_A, note that if S is the Riemann sphere (of diameter 1), $d\alpha$ is the element of area in the z-plane near the point z, and dA is the corresponding element of area on S, then

$$dA = \frac{d\alpha}{(1 + |z|^2)^2}.$$

Hence $\pi A(r)$ is exactly the area (counting multiplicity) of the image on the Riemann sphere of $\{|z| < r\}$ by $w = f(z)$. The rotation described above leaves $A(r)$ invariant and replaces $M_A(r, \infty)$, $N(r, \infty)$ by $M_A(r, a)$, $N(r, a)$. We see that $T_A(r, \infty)$ can be interpreted as an average of the spherical area of the image of disks under the mapping $w = f(z)$.

7
The Poisson-Jensen Formula

The material we present in this chapter is a specialization of some general results of potential theory. Our presentation is in the context of analytic function theory.

We consider functions f holomorphic in $\overline{\mathbb{D}}_R = \{z \in \mathbb{C} : |z| \leq R\}$. We denote $u = \operatorname{Re} f$, choose $r < R$, and write $z = re^{i\theta}$, $w = Re^{i\varphi}$.

The Poisson Formula.

$$(7.1) \quad u(re^{i\theta}) = \int P_r u = \frac{1}{2\pi} \int_{-\pi}^{\pi} u(Re^{i\varphi}) \frac{R^2 - r^2}{R^2 - 2rR\cos(\theta - \varphi) + r^2} \, d\varphi.$$

The Poisson Kernel.

$$(7.2) \quad \begin{aligned} P = P(z, w) = P(re^{i\theta}, Re^{i\varphi}) &= \frac{R^2 - r^2}{R^2 - 2rR\cos(\theta - \varphi) + r^2} \\ &= \frac{|w|^2 - |z|^2}{|w - z|^2} = \operatorname{Re} \frac{w + z}{w - z} \end{aligned}$$

Proof. Without loss of generality, assume $R = 1$:

$$\frac{1}{w - z} + \frac{\bar{z}}{1 - \bar{z}w} = \frac{1 - |z|^2}{(w - z)(1 - \bar{z}w)} = \frac{1}{w} \frac{1 - |z|^2}{|w - z|^2}.$$

By the Cauchy integral formula,

$$\frac{1}{2\pi i} \int_{|w|=1} f(w) \left[\frac{1}{w - z} + \frac{\bar{z}}{1 - \bar{z}w} \right] dw = f(z).$$

After parametrizing the integral with respect to the angle φ and taking real parts, we get the Poisson formula.

Remark. For $z = 0$, Poisson's formula reduces to the Gauss Mean Value Theorem,

$$u(0) = \frac{1}{2\pi} \int_{-\pi}^{\pi} u(e^{i\varphi}) \, d\varphi.$$

The Poisson formula is the "invariant form" of the Gauss Mean Value Theorem in the following sense. Choose $z \in \mathbb{D}$ and define

$$T_z : \overline{\mathbb{D}} \to \overline{\mathbb{D}} \quad \text{by} \quad T_z w = \frac{w + z}{1 + \bar{z}w} \quad \text{for} \quad w \in \mathbb{D}.$$

Let

$$F = f \circ T_z, \quad U = u \circ T_z.$$

If $\lambda = \frac{w+z}{1-\bar{z}w}$, then $w = \frac{\lambda - z}{1 - \bar{z}w}$ and

$$\frac{dw}{w} = \frac{1 - \bar{z}w}{|\lambda - z|^2} \frac{d\lambda}{\lambda},$$

so that

$$u(z) = U(0) = \frac{1}{2\pi i} \int U(w) \frac{dw}{w} = \frac{1}{2\pi i} \int u(\lambda) \frac{1 - |z|^2}{|\lambda - z|^2} \frac{d\lambda}{\lambda},$$

which is the Poisson formula in different notation.

This generalization of the Gauss Mean Value Theorem has a natural application which generalizes Jensen's Theorem.

The Poisson-Jensen Formula.
Suppose that f is meromorphic in the disk $\mathbb{D}_R = \{z \in \mathbb{C} : |z| < R\}$, $r < R$. Then,
(7.3)

$$\log|f(re^{i\theta})| = \frac{1}{2\pi} \int_{-\pi}^{\pi} \log|f(Re^{i\theta})| \frac{R^2 - r^2}{R^2 - 2rR\cos(\theta - \varphi) + r^2} \, d\varphi$$

$$+ \sum_{|z_\nu| < R} \log|B_R(z : z_\nu)| - \sum_{|w_\nu| < R} \log|B_R(z : w_\nu)| - k \log \frac{R}{r},$$

where B is the Blaschke factor defined by

$$B_R(z : a) = \frac{R(z - a)}{R^2 - \bar{a}z}$$

and the z_ν are the zeros of f, the w_ν are the poles of f, and k is the order of the zero or pole at the origin.

Corollary. *If f is holomorphic, then*

$$\log |f(re^{i\theta})| \le \int_{|w|=R} P \log |f|.$$

In other words, if f is holomorphic, then $\log |f|$ is dominated by its Poisson integral. Notice that for $z = 0$ the Poisson-Jensen formula reduces to Jensen's Theorem. One way to prove the Poisson-Jensen formula is to show that it is the invariant form of Jensen's Theorem. We choose another proof.

Proof of the Poisson-Jensen Formula. If f is holomorphic and has no zeros, then there is a branch F of $\log f$ and $\log |f| = Re F$ so that the formula follows as a special case of the Poisson formula. Also, if λ denotes the left-hand side and ρ the right-hand side, notice that $\lambda(fg) = \lambda(f) + \lambda(g)$ and $\rho(fg) = \rho(f) + \rho(g)$. So it is enough to prove the formula for holomorphic f. Now consider $g(z) = f(z)/\Pi B_R(z : z_\nu)$ supposing that $f(0) \neq 0$. Since $|B_R(w : z_\nu)| = 1$ for $|w| = R$, the formula follows on applying the Poisson formula to g. And if $f(0) = 0$, consideration of $f(z)/z^k$ leads to the general case.

8
Applications of $T(r)$

Theorem. *If f is holomorphic and $M(r) = \sup[\{|f(z)| : |z| \le r\}$, then for any $R > r$*

$$T(r) \le \log^+ M(r) \le \frac{R+r}{R-r} T(R).$$

Proof. Since f is holomorphic, $T(r) = m(r)$ and

$$m(r) = \frac{1}{2\pi} \int_{-\pi}^{\pi} \log^+ |f(re^{i\theta})| \, d\theta \le \log^+ M(r).$$

Also, by the corollary to the Poisson-Jensen formula,

$$\log |f(re^{i\theta})| \le \frac{1}{2\pi} \int_{|z|=R} P \log |f|.$$

It is easy to verify that

$$0 < P \le \frac{R+r}{R-r}$$

so that

$$\log^+ M(r) \le \frac{R+r}{R-r} m(r) = \frac{R+r}{R-r} T(R),$$

and the theorem is proved.

We now can prove the following extension of the Liouville Theorem, which is left as an exercise.

Exercise. Suppose that f is meromorphic in $|z| < \infty$ and that $T(r)$ is bounded. Then f is a constant.

Definition. Suppose that $\lambda(r)$ is positive, continuous, and nondecreasing for $r > 1$ and furthermore is slowly increasing in the sense that $\frac{\lambda(2r)}{\lambda(r)}$ is bounded. Let Λ^* be the class of all entire functions f such that

$$\log^+ M(r) = O(\lambda(r)).$$

It is easy to see that Λ^* is a ring. One of our exercises is to show that $f \in \Lambda^*$ if and only if $f \in \ell$, where ℓ consists of those entire functions f for which $T(r, f) = O(\lambda(r))$. In the case where $\lambda(r) = \max(1, r^\rho)$, we see that the notions of f being of finite order, of order at most ρ, and of order at most ρ exponential type are the same whether defined by the characteristic or the logarithm of the maximum modulus. It also follows that each Λ^* ring is algebraically closed in the ring of all entire functions.

We say that a meromorphic function f in the unit disc is of bounded characteristic to mean that $T(r, f)$ is bounded. Next we characterize the functions of a bounded characteristic in the unit disk $\mathbb{D}$.

Theorem. *A function f meromorphic in $\mathbb{D}$ is of bounded characteristic if and only if there exist bounded functions A and B, holomorphic in $\mathbb{D}$, such that $f = A/B$.*

Proof. It is easy to see that if $f = A/B$, then f is of bounded characteristic.

In the other direction, suppose that f is of bounded characteristic and, without loss of generality, suppose $f(0) = 1$. Since $T(r, f)$ is bounded, it follows that $N(r, f)$ and $N(r, \frac{1}{f})$ are bounded, so that

$$\sum \log \frac{1}{r_n} < \infty \quad \text{and} \quad \sum \log \frac{1}{\rho_n} < \infty.$$

Here, as before, $\{r_n e^{i\theta_n}\}$ and $\{\rho_n e^{i\varphi_n}\}$ are the zeros and poles of f.

By the theorem of Chapter 4, there exist bounded functions φ and ψ, holomorphic in $\mathbb{D}$, such that the zeros of φ are the zeros of f and the zeros of ψ are the poles of f. Thus, $g = \frac{f\psi}{\varphi}$ is of bounded characteristic and has no zeros or poles.

It is enough to show that g has a representation $g = A/B$. Note that even though g is holomorphic with no zeros and $T(r, g) = O(1)$, it does not follow that g is bounded; witness $g(z) = \frac{1}{1-z}$.

Proceeding with the proof, there exists a function h, holomorphic in $\mathbb{D}$, such that $g = e^h$. Writing $h = u + iv$, we have

$$\frac{1}{2\pi} \int_{-\pi}^{\pi} |u(re^{i\theta})| \, d\theta \leq M < \infty \quad \text{for} \quad r < 1,$$

since $m(r, g)$ and $m(r, \frac{1}{g})$ are bounded, and $|g| = \exp u$. Now for $R < 1$, writing

$$h_R(z) = h(Rz) = u_R(z) + iv_R(z),$$

we have

$$\frac{1}{2\pi} \int_{-\pi}^{\pi} |u_R(re^{i\theta})| \, d\theta \leq M \quad \text{for} \quad R \leq 1.$$

Now we write

$$u_R = u_R^+ - u_R^-; \quad u_R^+ = \max(0, u_R).$$

Let

$$\beta_R(z) = \frac{1}{2\pi} \int_{-\pi}^{\pi} u_R^+(e^{i\varphi}) \frac{e^{i\varphi} + z}{e^{i\varphi} - z} \, d\varphi$$

$$\alpha_R(z) = \frac{1}{2\pi} \int_{-\pi}^{\pi} u_R^-(e^{i\varphi}) \frac{e^{i\varphi} + z}{e^{i\varphi} - z} \, d\varphi.$$

It is easy to verify that α_R and β_R are holomorphic in $\mathbb{D}$. Let

$$g_R = \exp\left(i\lambda_R \frac{A_R}{B_R}\right),$$

where

$$A_R = \exp(-\alpha_R),$$
$$B_R = \exp(-\beta_R),$$

and λ_R is an appropriate real constant. Now

$$Re\,\alpha_R \geq 0 \quad \text{and} \quad Re\,\beta_R \geq 0$$

so that

$$|A_R| \leq 1 \quad \text{and} \quad |B_R| \leq 1.$$

Since the families $\{A_R\}$ and $\{B_R\}$ are uniformly bounded, they are normal families; and since the unit circumference is compact, we may write, for a suitable sequence of R approaching 1,

$$\lim A_R = A, \quad \lim B_R = B, \quad |A| \leq 1, \quad |B| \leq 1, \quad \lim \lambda_R = \lambda.$$

It is easy to see that

$$\lim g_R = g.$$

We therefore get the required representation

$$g = e^{i\lambda} \frac{A}{B}$$

provided only that B is not identically zero. But

$$|B_R(0)| = \exp(-\beta_R(0))$$

and

$$\beta_R(0) = \frac{1}{2\pi} \int_{-\pi}^{\pi} u_R^+(e^{i\theta}) \, d\theta \leq M.$$

The proof is complete.

9
A Lemma of Borel and Some Applications

Definition. A set E of real numbers has length $\leq \ell$, written $|E| \leq \ell$, means that there is a countable union of intervals $[a_n, b_n]$, $a_n \leq b_n$, that contains E and such that

$$\sum(b_n - a_n) \leq \ell.$$

Definition. $|E| = \inf\{\ell : |E| \leq \ell\}$.

Lemma (trivial). *If $|E_1| \leq \ell_1$ and $|E_2| \leq \ell_2$, then*

$$|E_1 \cup E_2| \leq \ell_1 + \ell_2.$$

Borel Lemma. *Suppose that $\mu(r)$ is defined for all $r \geq r_0$, that μ is nondecreasing, and that $\mu(r_0) \geq 1$. Then for each $a > 1$*

(9.1)
$$\mu\left(r + \frac{1}{\mu(r)}\right) < a\mu(r)$$

except in a set E_a such that $|E_a| \leq \frac{a}{a-1}$.

Remark. The inequality of the Borel Lemma estimates μ at a point greater than r by using the value that μ takes at r. Intuitively, if the inequality fails in "too big" a set, the function μ will become infinite "too soon" and will not be defined for all r.

Proof. Let $E = E_a = \{r > r_0 : \mu\left(r + \frac{1}{\mu(r)}\right) \geq a\mu(r)\}$. Let $\epsilon > 0$ be given. We proceed to define a sequence $\{r_n\}$ and an allied sequence $\{r'_n\}$ by induction.

Let $r_1 = \inf\{r : r \in E\}$. Suppose $r_1, \ldots, r_{n-1}$ have been constructed together with numbers $\epsilon_k \geq 0$, $k = 1, \ldots, n-1$ so that $\epsilon_1 + \epsilon_2 + \ldots \epsilon_{n-1} < \epsilon$ and $r_k + \epsilon_k \in E$. Let $r'_k = r_k + \epsilon_k + \frac{1}{\mu(r_k + \epsilon_k)}$.

Now define $r_n = \inf\{r : r \in E \text{ and } r \geq r'_{n-1}\}$. Choose $\epsilon_n \geq 0$ so that $\epsilon_1 + \cdots + \epsilon_n < \epsilon$ and so that $r_n + \epsilon_n \in E$ and proceed.

This procedure will terminate after n steps if and only if there does not exist an $r \in E$ satisfying $r \geq r'_n$. We have

(9.2) $$r_n \leq r_n + \epsilon_n < r'_n \leq r_{n+1}.$$

Now $(r'_n, r_{n+1}) \cap E$ is empty by construction.

Claim. There exist only finitely many r_n or else $r_n \to \infty$. For otherwise, there would be a finite r such that $r_n \to r$ and, by (10.2), $r'_n \to r$. However, by the construction we have for all k

$$r'_k - r_k = \epsilon_k + \frac{1}{\mu(r_k + \epsilon_k)} \geq \frac{1}{\mu(r)} > 0,$$

which is a contradiction.

Claim. $E \subset \bigcup_{n=1}^{\infty}[r_n, r'_n]$.

Pick an arbitrary $x \in E$. Let $r_{n_0} = \max[r_n : r_n \leq x]$. This makes sense by the previous claim. Now $r_{n_0} \leq x$. Suppose $x > r'_{n_0}$. Then $r_{n_0+1} \leq x$, an immediate contradiction. Therefore $r_{n_0} \leq x \leq r'_{n_0}$, which proves the claim.

So we have constructed a countable collection of intervals whose union contains the set E. We estimate $\sum(r'_n - r_n)$. Notice that

$$\mu(r'_n) = \mu\left(r_n + \epsilon_n + \frac{1}{\mu(r_n + \epsilon_n)}\right) \geq a\mu(r_n + \epsilon_n) \geq a\mu(r_n)$$

so that $\mu(r_{n+1}) \geq a\mu(r_n)$. Therefore, $\mu(r_{n+1}) \geq a^n\mu(r_1) \geq a^n$. Hence

$$\sum(r'_n - r_n) = \sum\left(\epsilon_n + \frac{1}{\mu(r_n + \epsilon_n)}\right) \leq \epsilon + \sum\frac{1}{\mu(r_n + \epsilon_n)}.$$

But

$$\sum\frac{1}{\mu(r_n + \epsilon_n)} \leq \sum\frac{1}{\mu(r_n)} \leq \sum_{n=1}^{\infty}\frac{1}{a^{n-1}} = \frac{a}{a-1}.$$

It now follows that $|E| \leq \frac{a}{a-1} + \epsilon$, and the lemma is proved.

Corollary. *Under the same hypotheses on μ and a,*

$$(9.3) \qquad \mu\left(r\left(1 + \frac{1}{\mu(r)}\right)\right) < a\mu(r)$$

except for r in a set E_a', where E_a' has logarithmic length $\leq \frac{a}{a-1}$.

(By this we mean that $E_a' = \exp E_a$, where $|E_a| \leq \ell$. We write $|E_a'|_{\log} = |E_a|$.)

Proof. Let $\mu_1(y) = \mu(\exp y)$. Then $\mu\left(\exp\left(y + \frac{1}{\mu(y)}\right)\right) \leq a\mu(\exp y)$ for $y \notin E_a$, where $|E_a| \leq \frac{a}{a-1}$ by the Borel Lemma. But

$$\exp\frac{1}{\mu(\exp y)} \geq 1 + \frac{1}{\mu(\exp y)}$$

so that

$$\mu\left(\left\{1 + \frac{1}{\mu(\exp y)}\right\}\exp y\right) < a\mu(\exp y) \quad \text{for} \quad y \notin E_a,$$

and the result follows on writing $r = \exp y$.

Application of the Borel Lemma to Nevanlinna Theory

We already have proved that if f is entire, then

$$T(r) \leq \log M(r) \leq \frac{R+r}{R-r}T(R) \quad \text{if} \quad R > r.$$

Choose

$$R = r\left(1 + \frac{1}{T(r)}\right)$$

to get

$$\log M(r) \leq (2T(r) + 1)T\left(r\left(1 + \frac{1}{T(r)}\right)\right).$$

Unless f is a constant, $T(r) \to \infty$ so that $2T(r) + 1 \leq \frac{5}{2}T(r)$ for large r. Applying the Borel Lemma we find for entire functions f,

$$(9.4) \qquad \log M(r) \leq 3(T(r))^2$$

except for a set of finite logarithmic length.

Definition. We say that $A(r) \underset{\text{eff}}{\to} L$ means that there is a set E of finite logarithmic length such that $\lim_{r \notin E} A(r) = L$ as $r \to \infty$.

We attach a similar meaning to expressions like $A(r) \underset{\text{eff}}{\sim} B(r)$, $A(r) \underset{\text{eff}}{=} B(r)$, etc.

We have, in effect, proved the next result.

Proposition. *If f is a nonconstant entire function, then*

$$\log \log \log M(r) \underset{\text{eff}}{\sim} \log \log T(r).$$

In a certain sense, this says that $T(r)$ and $\log M(r)$ have the same size for most of the r. The $\log \log$ takes a lot of punishment.

Propostion. *Suppose that a real function f has a continuous, increasing derivative on $[1, \infty]$, and that $\lim_{x \to \infty} f(x) = \infty$. Then*

$$\log \log(f(x)) \underset{\text{eff}}{\sim} \log \log(x f'(x)).$$

Proof. Write $\nu(t) = t f'(t)$, and suppose, without loss of generality, that $f(1) = 0$. Then

$$f(x) = \int_1^x f'(t) \, dt = \int_1^x \nu(t) \frac{dt}{t}.$$

Hence $f(x) \le \nu(x) \log x$. But for some $\epsilon > 0$, we must have $\nu(x) \ge \epsilon x$ for x large, so that

(9.5) $$f(x) \le (\nu(x))^2 \quad \text{for large} \quad x.$$

In the other direction, if $y < x$, then

$$f(x) \ge \int_y^x \nu(t) \frac{dt}{t} \ge \nu(y) \log\left(\frac{x}{y}\right).$$

Choose $x = y + \frac{y}{f(y)}$ to get

$$\nu(y) \le f\left(y + \frac{y}{f(y)}\right) \frac{1}{\log\left(1 + \frac{1}{f(y)}\right)}.$$

Since $\log(1 + t) \ge \frac{1}{2} t$ for t near 0 and positive, we have, if y is large, $\nu(y) \le 2 f(y) f\left(y + \frac{y}{f(y)}\right)$. Applying the corollary of the Borel Lemma, we get

(9.6) $$\nu(y) \underset{\text{eff}}{\le} e(f(x))^2.$$

Equations (9.5) and (9.6) together imply the result.

10
The Maximum Term of an Entire Function

We will give in this chapter a proof that a suitable entire function can grow as fast as we please.

Let $f(z) = \sum a_n z^n$ be an entire function; $a_0 \neq 0$

$$A_n = |a_n|$$
$$|z| = r.$$

For each r, the sequence $A_0, A_1 r, A_2 r^2, \ldots$ converges to zero. Therefore we can define

(10.1) $$B(r) = \max(A_0, A_1 r, A_2 r^2, \ldots).$$

$B(r)$ is called the *maximum term* for r. A term $A_k r^k$ is a maximum term if $A_k r^k = B(r)$.

Since each $A_k r^k$ is a nondecreasing function of r (increasing if $k \neq 0$, $A_k \neq 0$), $B(r)$ is nondecreasing. $B(r)$ is also continuous and unbounded.

We define the *rank of the maximum term* as

(10.2) $$\mu(r) = \sup(n : A_n r^n = B(r)).$$

It follows immediately that if $n < \mu(r)$, then $A_n r^n \leq A_{\mu(r)} r^{\mu(r)}$, and if $n > \mu(r)$, then $A_n r^n < A_{\mu(r)} r^{\mu(r)}$. Therefore we also can write the rank of the maximum term as

$$\mu(r) = \sup(n : A_n r^n \geq B(r)).$$

Clearly, $\mu(r)$ is a nondecreasing, integer-valued function of r. Let

(10.3)
$$\begin{cases} g_n = -\log A_n & \text{if } A_n \neq 0 \\ g_n = \infty & \text{if } A_n = 0. \end{cases}$$

Since f is entire, we have

(10.4)
$$\lim_{n \to \infty} A_n^{-\frac{1}{n}} = \infty \quad \text{so that} \quad \lim_{n \to \infty} \frac{g_n}{n} = \infty.$$

Let c_n be the point $(n, g(n))$ on the plane. From (10.4) it follows that below any straight line of finite slope there is only a finite number of points c_n.

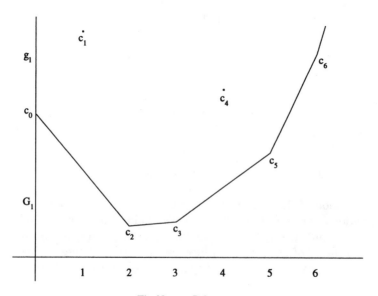

The Newton Polygon

This property of the c_n enables us to construct the *Newton polygon* $\pi(f)$ of the entire function f.

We construct the polygon as follows:

Among the segments $\overline{c_0 c_n}$, consider those of minimal slope. From these segments of minimal slope choose the one that is the longest; denote this segment by $\overline{c_0 c_{k_1}}$. Repeat this selection procedure starting with the point c_{k_1} to obtain the point c_{k_2}, and so on.

The vertices of the polygon are $\gamma_0, \gamma_1, \dots \gamma_i, \dots$, where $\gamma_i = c_{k_i} = (k_i, g(k_i))$ for $i = 1, 2, \dots$ and $\gamma_0 = c_{k_0} = c_0$. The x-coordinates of the vertices $\gamma_0, \gamma_1, \dots$ are called the *principal indices*.

Let G_n be the y-coordinate of the point on $\pi(f)$ whose x-coordinate is n. Let

(10.5)
$$A'_n = \exp(-G_n).$$

A'_n is called the *logarithmic convexification* of A_n. Since $A_n = \exp(-g_n)$, it follows immediately that

$$\begin{cases} G_n = g_n & \text{if } n \text{ is a principal index} \\ G_n \leq g_n & \text{for all } n. \end{cases}$$

Given $r > 0$, we have

$$\log A_\rho r^\rho \leq \log B(r) = \log A_n r^n, \quad \text{where} \quad n = \mu(r).$$

Since

$$\begin{cases} \log A_\rho = -g_\rho \\ \log A_n = -g_n, \end{cases}$$

we have

$$\rho \log r - g_\rho \leq n \log r - g_n$$

or

(10.6) $$g_\rho \geq g_n + (\rho - n) \log r.$$

But

(10.7) $$y = g_n + (x - n) \log r$$

is the equation of the straight line through c_n with slope $\log r$.

Equation (10.6) says that all points of $\pi(f)$ lie above the line (10.7); this line is "tangent" to $\pi(f)$. Call this line D_r. Thus, $\mu(r)$ is the rightmost point of contact of D_r with $\pi(f)$ [because $\mu(r)$ is the largest value of n at which we can have equality in (10.6)]. Hence the values of $\mu(r)$ are the principal indices.

Since, for $x = 0$, (10.7) yields $y = g_n - n \log r = -\log A_n r^n = -\log B(r)$, it follows that D_r cuts the y axis at $-\log B(r)$. Two immediate consequences of this are:

(a) Given f_1, f_2 entire such that $\pi(f_1) = \pi(f_2)$, then

$$\begin{cases} \mu(r : f_1) = \mu(r : f_2) \\ B(r : f_1) = B(r : f_2) \end{cases}$$

(b) Among all entire functions, $h(z) = \sum A'_n z^n$ is the largest function that has the same μ and B as f.

Let

(10.8) $$R_n = e^{G_n - G_{n-1}} = \frac{-A'_{n-1}}{A'_n}.$$

We call the R_n the *corrected ratios*.

Geometrically, $\log R_n$ is the slope of the side of $\pi(f)$ joining the points whose x-coordinates are $n-1$ and n. R_n is nondecreasing and $R_n \to \infty$ as $n \to \infty$.

Without loss of generality, assume that $A_0 = a_0 = 1$ [note that $\mu(0) = 0$]. From the definition of R_n, it follows that $e^{G_n} = R_1 \cdot R_2 \cdots R_n$.

Since $\mu(r)$ runs through the principal indices, $g_{\mu(r)} = G_{\mu(r)}$, it follows that

$$B(r) = \frac{r^{\mu(r)}}{e^{G_n}} = \frac{r^{\mu(r)}}{R_1 \cdot R_2 \cdots R_n}.$$

Taking logarithms,

$$\log B(r) = \mu(r)\log r + \sum_1^{\mu(r)} \log \frac{1}{R_k} = \sum_1^{\mu(r)} \log \frac{r}{R_k}.$$

But

$$\sum_1^{\mu(r)} \log \frac{r}{R_k} = \int_0^r \log \frac{r}{t}\, d\mu(t), \quad \text{where} \quad \mu(t) = \sum_{R_k \leq t} 1.$$

Integrating by parts we get:

$$\int_0^r \log \frac{r}{t} d\mu(t) = \log \frac{r}{t} \cdot \mu(t) \Big|_{0+}^r + \int_0^r \frac{\mu(t)}{t}\, dt$$

or

(10.9)
$$\log B(r) = \int_0^r \frac{\mu(t)}{t}\, dt,$$

and by the lemma in Chapter 6 it follows that $B(r)$ is a convex function of $\log r$.

Relation between $B(r)$ and $M(r)$
From Cauchy's inequality, $|a_n| \leq r^{-n} M(r)$, we get

(10.10)
$$|a_n|r^n \leq M(r) \quad \text{or}$$
$$B(r) \leq M(r).$$

Remark. $M(r) \leq F(r) \equiv \sum A'_n r^n$.
Note: Choose $p > \mu(r)$. Then $R_p > r$ because $\log R_p > \log r$ for $p > \mu(r)$, as we see on interpreting $\log R_p$ as a slope. Then for $q \geq p$ we can write

$$A'_q r^q = e^{-G_q} r^q = e^{G_{p-1}} r^{p-1} \frac{r^{q-p+1}}{R_p \cdots R_q}.$$

Since $e^{-G_{p-1}} r^{p-1} < B(r)$, we have

$$A'_q r^q = B(r)\frac{r^{q-p+1}}{R_\rho^{q-p+1}} \leq B(r)\left(\frac{r}{R_p}\right)^{q-p+1}$$

because the slopes of the edges of the Newton polygon are increasing. We have:

$$F(r) = \sum_{0}^{p-1} e^{-G_n} r^n + \sum_{p}^{\infty} e^{-G_n} r^n$$

$$\leq pB(r) + B(r) \sum_{q=p}^{\infty} \left(\frac{r}{R_p}\right)^{q-p+1}$$

$$= pB(r) + B(r) \frac{r}{R_p - r}.$$

Consequently,

$$F(r) \leq B(r) \left[p + \frac{r}{R_p - r}\right] \quad \text{provided} \quad p > \mu(r).$$

As a heuristic guide, let us try the choice $p = \mu(x)$ for $x > r$, supposing for the moment that $\mu(x) > \mu(r)$. Then

$$F(r) \leq B(r) \left[\mu(x) + \frac{r}{R_{\mu(x)} - r}\right] \leq B(r) \left[\mu(x) + \frac{r}{R_{x-r}}\right].$$

The last inequality is justified as follows: $R_{\mu(x)}$ is the slope between $(p - 1, G_{p-1})$ and (p, G_p). This slope grows without bound, so there exists an x so that for $p > x$ we have the inequality. Write $x = r + y$; $y > 0$. Then

$$F(r) \leq B(r) \left[\mu(r + y) + \frac{r}{y}\right].$$

Now write $y = \frac{r}{t}$ so that

$$F(r) \leq B(r) \left[\mu\left(r + \frac{r}{t}\right) + t\right].$$

We try to make

$$t = \mu\left(r + \frac{r}{t}\right) = \mu\left(r + \frac{r}{\mu\left(r + \frac{r}{\mu(r + \frac{r}{t})}\right)}\right) = \cdots$$

by choosing t.

As a first approximation we choose $t = \mu(r)$. For that choice $x = r + \frac{r}{\mu(r)}$. Since we want to guarantee that $p > \mu(r)$, our actual choice is $p = \mu\left(r + \frac{r}{\mu(r)}\right) + 1$. We then get:

$$F(r) \leq B(r) \left[\mu\left(r + \frac{r}{\mu(r)}\right) + \mu(r) + 1\right]$$

or

$$F(r) \leq B(r)\left[2\mu\left(r + \frac{r}{\mu(r)}\right) + 1\right].$$

Using the Borel Corollary (Chapter 9) we find

$$F(r) \underset{\text{eff}}{\leq} 3B(r)\mu(r)$$

so that

(10.11) $$B(r) \leq M(r) \underset{\text{eff}}{\leq} 3B(r)\mu(r).$$

For functions of finite order, say order ρ, we have

$$\log M(r) = o(r^{\rho'}) \quad \text{as} \quad r \to \infty \quad \text{if} \quad \rho' > \rho.$$

In this case,

$$\log B(r) = o(r^{\rho'}) \quad \text{as} \quad r \to \infty \quad \text{if} \quad \rho' > \rho.$$

We have

$$\log B(r) = \int_0^r \frac{\mu(t)}{t}\, dt = o(r^{\rho'})$$

and we know that

$$\mu(r)\log 2 \leq \int_r^{2r} \frac{\mu(t)\, dt}{t} \leq \int_0^{2r} \frac{\mu(t)\, dt}{t} = o(r^\rho).$$

Hence,

(10.12) $$\mu(r) = o(r^\rho).$$

Together with (10.11), this implies that if either $\log M$ or $\log B$ is $o(r^{\tilde{\rho}})$, $\tilde{\rho} \leq \rho$, then both are. Now, from (10.11) we have

$$\log M(r) \underset{\text{eff}}{\leq} \log 3 + \log B(r) + \log \mu(r)$$
$$= \log 3 + o(r^\rho) + \rho \log r.$$

From $B(r) \leq M(r)$ we get

(10.13) $$\limsup_{r \to \infty} \frac{\log B(r)}{\log M(r)} \leq 1.$$

Also,

$$\frac{\log M(r)}{\log B(r)} \underset{\text{eff}}{\leq} \frac{\log 3 + \log B(r) + \log \mu(r)}{\log B(r)} \leq 1 + \frac{\log 3 + \log \mu(r)}{\log B(r)}.$$

Writing $F(r)$ as

$$F(r) = \sum A'_n r^n = \sum A'_n R^n \left(\frac{r}{R}\right)^n$$

and using the definition of $B(r)$, we get for $R > r$:

$$F(r) \leq B(R) \sum \left(\frac{r}{R}\right)^n = B(r)\frac{R}{R-r}$$

so that

(10.14) $$B(R) \leq M(r) \leq B(R)\frac{R}{R-r}.$$

Let us choose $R = r + \frac{r}{B(r)}$ and again apply the Borel Corollary (9.3):

$$B(r) \leq M(r) \leq B\left(r + \frac{r}{B(r)}\right)\frac{r + \frac{r}{B(r)}}{\frac{r}{B(r)}}$$

$$B(r) \leq M(r) \leq B\left(r + \frac{r}{B(r)}\right)(B(r) + 1).$$

Hence,

$$B(r) \leq M(r) \underset{\text{eff}}{\leq} 2B(r)^2.$$

Taking logarithms twice we obtain

(10.15) $$\log\log M(r) \underset{\text{eff}}{\sim} \log\log B(r).$$

From (10.11) we get

$$B(r) \leq M(r) \leq B(r) \cdot 3 \cdot \mu\left(r + \frac{r}{\mu(r)}\right).$$

If $\mu(r) = o(r^\rho)$, $\rho < \infty$, then $\mu\left(r + \frac{r}{\mu(r)}\right) \leq \mu(2r) = o(r^\rho)$ so that

$$\log B(r) \leq \log M(r) \leq \log B(r) + \rho\log r + \text{constant}.$$

Then

$$\frac{\log B(r)}{\log M(r)} \leq 1 \leq \frac{\log B(r)}{\log M(r)} + \frac{\rho\log r}{\log M(r)}.$$

We shall prove that $\frac{\rho\log r}{\log M(r)} = o(1)$, which implies:

Theorem 10.1. *If f is of finite order, then*

$$\log B(r) \sim \log M(r).$$

Assertion: $\frac{\rho \log r}{\log M(r)} = o(1)$ (unless f is a polynomial).

Proof. Otherwise, we can find a sequence $\{r_n\}$ such that $r_n \to \infty$ and such that
$\log M(r_n) \le c \log r_n$ or $M(r_n) \le r_n^c$. But Cauchy's inequality says that

$$|a_k| \le \frac{M(r_n)}{r_n^k} \le \frac{r_n^c}{r_n^k}.$$

It follows that $a_k = 0$ for $k > c$, and hence f is a polynomial.

In general, we have:

Theorem 10.2. *If $\liminf_{r \to \infty} \frac{T(r)}{\log r} = c < \infty$, then f is a rational function.*

Proof. We again find an increasing sequence $\{r_n\}$, $r_n \to \infty$ such that

$$T(r_n) \le c \log r_n$$

so that

$$N(r_n, f) \le c \log r_n.$$

We may suppose without loss of generality that $f(0) \ne 0, \infty$. Then:

$$\int_0^{r_n} \frac{n(t, f)}{t}\, dt \le c \log r_n$$

and for $s > 0 : \int_s^{r_n} \frac{n(t,f)}{t}\, dt \le c \log r_n$. Then

$$n(s, f) \log \frac{r_n}{s} \le c \log r_n \quad \text{or}$$

$$n(s, f) \le c \frac{\log r_n}{\log \frac{r_n}{s}}.$$

Let $r_n \to \infty$, and we see that $n(s, f) \le c$.

Therefore, f has at most c poles $\alpha_1, \ldots, \alpha_k$ with $k \le c$. Now, multiplying f by
$(z - \alpha_1) \cdots (z - \alpha_k)$ and applying the previous result to the holomorphic function so obtained, we complete the proof of the theorem.

Suppose we are given $M'(r)$ such that $\log M'(r)$ is logarithmically convex. We shall assume that $M'(r)$ can be written in the form

$$M'(r) = \int_0^r \frac{\gamma(t)}{t}\, dt; \quad \gamma(t) \quad \text{increasing}.$$

We already have proved (in Chapter 7) that $\int_0^r \frac{\gamma(t)}{t}\, dt$, where $\gamma(t)$ is increasing, is logarithmically convex. Suppose further that $\log M'(r) = o(r^\rho)$ for some $\rho < \infty$. Then there exists an entire function f such that $\log M(r, f) \sim \log M'(r)$. Thus, for any such $M'(r)$ there is an entire function f whose maximum modulus grows essentially like $M'(r)$.

The idea of the proof is the following: Take $\mu(t) = \gamma(t)$. Define $B(r) = \int_0^r \frac{\mu(t)}{t}\, dt$ and draw a Newton polygon associated with μ and B. This gives the A_n'. Let f be the associated function $f = \sum A_n' z^n$. Then

$$\log M(r) \sim \log B(r) = \log M'(r).$$

Since $\mu(t)$ must be integer-valued, for the actual proof we take

$$\mu(t) = [\gamma(t)] \quad \text{where}$$
$$[x] = \quad \text{greatest integer not exceeding } x.$$

We have then

$$\gamma(t) - 1 \le \mu(t) \le \gamma(t).$$

Since $\log M'(r) = o(r^\rho)$, it follows that $\gamma = o(r^\rho)$ so that $\mu(r) \sim \log B(r)$. It remains to be shown that

$$\log B(r) = \log M'(r) + O(\log r).$$

But

$$-\log \frac{r}{r_0} \le \int_{r_0}^r \frac{\mu(t) - \gamma(t)}{t}\, dt \le 0$$

so that

$$|\log B(r) - \log M'(r)| = O(\log r).$$

Hence $\log M(r) \sim \log M'(r)$, and the proof is complete.

Dropping the hypothesis of finite order, we can still get the following result: Given that $\log M'(r)$ is logarithmically convex, it is possible to find an entire function f such that $M(r) \ge M'(r)$.

Proof.

$$\log B(r) \le \log M(r)$$
$$\log B(r) \ge \log M'(r) + O(\log r).$$

The term $O(\log r)$ is easily disposed of by multiplying by a suitable polynomial.

Another result along this line is the following:

Theorem 10.3.   *Given any continuous function $M'(r)$, there exists an entire function f such that $M(r) \geq M'(r)$.*

Proof. It is easy to see that any such function $M'(r)$ has an increasing majorant such that $\log M'(r)$ is logarithmically convex, and, indeed, $\log M'(r) = \int_0^r \frac{\gamma(t)}{t} \, dt$. Now proceed as in the previous theorem and the proof is complete. Thus, there exist entire functions that grow as rapidly as we please.

We conclude this chapter with some more estimates on $B(r)$. We know that:

$$(10.16) \qquad B(r) < F(r) < B(r) \left[r + \frac{r}{\mu(r)} + 1 \right]$$

and that for $R > r$

$$F(r) < B(r) \frac{R}{R - r}.$$

We can refine our results using the fact that

$$\log B(r) = \int_0^r \frac{\mu(t)}{t} \, dt.$$

Since $\int_r^R \frac{\mu(t)}{t} \, dt \leq \int_0^R \frac{\mu(t)}{t} \, dt$, we get

$$\log B(R) = \int_0^R \frac{\mu(t))}{t} \, dt \geq \int_r^R \frac{\mu(t)}{t} \, dt \geq \mu(r) \log \frac{R}{r}$$

so that

$$(10.17) \qquad \mu(r) \leq \frac{\log B(R)}{\log \frac{R}{r}}.$$

Choose $R = r + \frac{r}{\log B(R)}$. Then we have

$$\mu(r) \leq \frac{\log B \left(r + \frac{r}{\log B(r)} \right)}{\log \left(1 + \frac{1}{\log B(r)} \right)}$$

$$\leq 2 \log B \left(r + \frac{r}{\log B(r)} \right) \log B(r).$$

From the Borel Corollary it follows that

$$(10.18) \qquad \mu(r) \underset{\text{eff}}{\leq} 3 \log^2 B(r).$$

But then (10.11) gives

$$F(r) \leq 3 B(r) \log^2 B(r).$$

Hence,

$$\log F(r) \underset{\text{eff}}{\sim} \log B(r).$$

11
Relation Between the Growth of an Entire Function and the Size of Its Taylor Coefficients

Let F be an entire function and $M(r)$ be its maximum modulus for $|z| = r$. Suppose λ is a positive continuous increasing function for $r \geq 1$ such that $\frac{\lambda(2r)}{\lambda(r)}$ is bounded.

Definition. If $\log M(r) = O(\lambda(r))$, we say f is of *finite λ-type* and write $f \in \Lambda$

Proposition 11.1. *$\Sigma a_n z^n$ is of finite λ-type if and only if there exists a constant K such that $|a_n| \leq \frac{e^{K\lambda(r)}}{r^n}$ for each n.*

Proof. Suppose $\Sigma a_n z^n$ is of finite λ-type. Then, since $\log M(r) = O(\lambda(r))$, there exists a constant K such that $M(r) \leq e^{K\lambda(r)}$. Now the Cauchy inequality gives $|a_n| \leq \frac{M(r)}{r^n}$, which gives the result.

Conversely, suppose $|a_n| \leq \frac{e^{K\lambda(r)}}{(2r)^n} = \frac{e^{K\lambda(r)}}{2^n r^n}$ so that

$$|a_n| r^n \leq 2^{-n} e^{K\lambda(r)}.$$

Thus

$$\Sigma |a_n| r^n \leq e^{K\lambda(r)},$$

so that $M(r) \leq e^{K\lambda(r)}$ and $\log M(r) = O(\lambda(r))$.

Definition. An entire function f *is of order$\leq \rho$* if for each $\rho' > \rho$ there exist constants $A = A(\rho')$ and $K = K(\rho')$ such that

$$|f(z)| \leq Ae^{K|z|^{\rho'}} \quad \text{for all} \quad z.$$

Definition. An entire function f *is of order* ρ if it is of order $\leq \rho$ but not of order $\leq \rho_0$ for any $\rho_0 < \rho$.

We have the following facts immediately from the definitions:

$$\log M(r) \leq \log^+ A + Kr^{\rho'}$$
$$\log^+ \log^+ M(r) \leq \log^+ \log^+ A + \log^+ K + \rho' \log^+ r + \log^+ 2.$$

Thus we have $\frac{\log^+ \log^+ m(r)}{\log r} \leq \rho' + o(1)$ as $r \to \infty$.

Now let $\lambda = \limsup_{r \to \infty} \frac{\log^+ \log^+ M(r)}{\log r}$. Then by the above observation we have $\lambda \leq \rho'$ for all $\rho' > \rho$ and hence $\lambda \leq \rho$. In the other direction, we have $\frac{\log^+ \log^+ M(r)}{\log r} \leq \lambda + o(1)$ or

$$\log^+ \log^+ M(r) \leq (\lambda + o(1)) \log r.$$

Therefore,

$$M(r) \leq e^{e^{(\lambda + o(1)) \log r}} = e^{r^\lambda r^{o(1)}}.$$

Thus, for every $\lambda' > \lambda$ and r large ($r > r_0$) we have $M(r) \leq e^{r^{\lambda'}}$ or, more generally, for all r, $M(r) \leq Ae^{r^{\lambda'}}$. Hence f is of order $\leq \lambda'$ for all $\lambda' > \lambda$ and thus f is of order λ. We therefore have proved:

Proposition 11.2. *For any entire function f,*

$$\rho = \limsup_{r \to \infty} \frac{\log^+ \log^+ M(r)}{\log r}.$$

Definition. Suppose f is an entire function of order ρ and that $|f(z)| \leq Ae^{K|z|^\rho}$ for all z. Then we say f is of order ρ, *type at most* K. The *type* τ *of* f is the infimum of those numbers K such that f is of type at most K.

Definition. We say f is of *finite-type* if τ is finite; we say f is of *minimal-type* if $\tau = 0$; and we say f is of *mean-type* if it is of finite type but not of minimal-type.

Definition. We say f is of *growth* (ρ, τ) if either it is of order $< \rho$ or it is of order ρ and type $\leq \tau$. A function of growth $(1, \tau)$ is called *a function of exponential-type*.

Proposition 11.3. *For any entire function f of order ρ,*

$$\tau = \limsup_{r \to \infty} \frac{\log^+ M(r)}{r^\rho}.$$

The proof is straightforward.

In proving the following propositions we shall use the elementary fact that

$$\max_y \frac{R^y}{y^y} = \frac{R^{R/e}}{\left(\frac{R}{e}\right)^{R/e}} = e^{R/e}.$$

Proposition 11.4. *Given an entire function f, let*

$$\alpha = \limsup_{n \to \infty} \frac{n \log n}{\log \frac{1}{|a_n|}}.$$

Then $\alpha = \rho$.

Proof. Take $\sigma > \rho$. Then $|a_n| r^n \leq M(r) \leq e^{r^\sigma}$ for large r. Thus $|a_n| \leq r^{-n} e^{r^\sigma}$ or $\log |a_n| \leq r^\sigma - n \log r$. Now choose $r = \left(\frac{n}{\sigma}\right)^{\frac{1}{\sigma}}$, which is large for n large. Therefore we have

$$\log |a_n| \leq \frac{n}{\sigma} - \frac{n}{\sigma} \log \frac{n}{\sigma} \quad \text{or} \quad \log \frac{1}{|a_n|} \geq \frac{n}{\sigma} \log \frac{n}{\sigma} - \frac{n}{\sigma}.$$

Hence

$$\frac{n \log n}{\log \frac{1}{|a_n|}} \leq \frac{n \log n}{\frac{n}{\sigma} \log \frac{n}{\sigma} - \frac{n}{\sigma}} \leq \sigma + o(1) \quad \text{as} \quad n \to \infty.$$

Therefore $\alpha \leq \sigma$ and, since σ can be chosen arbitrarily close to ρ, $\alpha \leq \rho$.

Now take $\beta > \alpha$ so that $\frac{n \log n}{\log \frac{1}{|a_n|}} < \beta$ for large n and, without loss of generality, for all n. Thus $n \log n \leq \beta \log \frac{1}{|a_n|}$ or $n^{n/\beta} \leq \frac{1}{|a_n|}$ or $|a_n| \leq \frac{1}{n^{n/\beta}}$. Hence $|a_n| r^n \leq \frac{r^n}{n^{n/\beta}}$ and therefore $B(r) \leq \sup_x \frac{r^x}{x^{x/\beta}}$, where $B(r)$ is the maximum term of the series $\Sigma |a_n| r^n$ (see Chapter 10). If we let $y = \frac{x}{\beta}$ and $R = \frac{r}{\beta}$, we have $B(r^{1/\beta}) \leq \sup_x \frac{r^{x/\beta}}{x^{x/\beta}} = \sup \frac{r^y}{y^y \beta^y} = \sup_y \frac{R^y}{y^y} = e^{R/e} = e^{r/\beta e}$. Therefore $B(r) \leq \exp(r^\beta / e^{\beta e})$, $\log B(r) \leq \frac{r^\beta}{\beta e}$ and $\log \log B(r) \leq \beta \log r - \log \beta e$. Hence $\limsup_{r \to \infty} \frac{\log^+ \log^+ B(r)}{\log r} \leq \beta$. But $\log M(r) \sim \log B(r)$, and so we have $\rho \leq \beta$ and hence $\rho \leq \alpha$. We therefore must have $\rho = \alpha$.

Proposition 11.5. *If $f = \Sigma a_n z^n$ is of order $\leq \rho$, then*

$$\tau = \frac{1}{e\rho} \limsup_{n \to \infty} n |a_n|^{\rho/n}.$$

Proof. If f is of order ρ and type τ, take $\tau' > \tau$. Using the Cauchy inequality we have $|a_n| \leq \frac{M(r)}{r^n} \leq \frac{e^{\tau' r^\rho}}{r^n}$ for large r. We now minimize the expression on the right. Its logarithm is $\tau' r^\rho - n \log r$, and setting the derivative of this equal to zero, $\tau' \rho r^{\rho - 1} - \frac{n}{r} = 0$, we choose $r = \left(\frac{n}{\tau' \rho}\right)^{1/\rho}$. Thus $|a_n| \leq \frac{e^{n/\rho}}{\left(\frac{n}{\tau' \rho}\right)^{n/\rho}} = \left(\frac{e\tau' \rho}{n}\right)^{\frac{n}{\rho}}$. Hence $n |a_n|^{\rho/n} \leq e\tau' \rho$, so that we have $\limsup_{n \to \infty} n |a_n|^{\rho/n} \leq e\tau' \rho$ and therefore $\limsup_{n \to \infty} n |a_n|^{\rho/n} \leq e\tau \rho$.

Now take $\beta > \frac{1}{e\rho} \limsup_{n\to\infty} n|a_n|^{\rho/n}$. Then for large n we have $|a_n| \leq \left(\frac{e\beta\rho}{n}\right)^{n/\rho}$ so that $|a_n|r^n \leq \left(\frac{e\beta\rho}{n}\right)^{n/\rho} r^n$. Hence $B(r) \leq \left(\frac{e\beta\rho}{n}\right)^{n/\rho} r^n$ or $B(r) \leq \max_x \frac{(e\beta\rho)^{n/\rho}}{x^{x/\rho}} r^x$. If we let $y = \frac{x}{\rho}$ and $R = e\beta r$, we have

$$B(r^{1/\rho}) \leq \max_y \frac{(e\beta\rho)^y r^y}{\rho^y y^y} = \max_y \left(\frac{e\beta r}{y}\right)^y = \max_y \frac{R^y}{y^y} = e^{R/e} = e^{\beta r}.$$

Hence $B(r) \leq e^{\beta r^\rho}$ and thus $\limsup_{r\to\infty} \frac{\log B(r)}{r^\rho} \leq \beta$. But $\log B(r) \sim \log M(r)$ so that $\tau \leq \beta$ and thus $\tau \leq \frac{1}{e\rho} \limsup_{n\to\infty} n|a_n|^{\rho/n}$, which completes the proof.

Proposition 11.6.   *The orders and types of an entire function f and of its derivative f' are the same.*

This follows easily from the formulas for order and type in terms of the power series coefficients.

Corollary to Proposition 11.5.   *Write $f(z) = \Sigma \frac{\alpha_n}{n!} z^n$. Then f is of exponential-type if and only if $\Sigma \alpha_n \zeta^n$ has a finite radius of convergence.*

Proof. We use Stirling's formula $n! \sim n^n e^{-n}\sqrt{2\pi n}$.

Now we have $n\left|\frac{\alpha_n}{n!}\right|^{1/n} \sim n\left|\frac{\alpha_n}{n^n e^{-n}}\right|^{1/n} \sim e|\alpha_n|^{1/n}$. Furthermore, the radius of convergence of $\Sigma \alpha_n \zeta^n$ is $\frac{1}{\limsup_{n\to\infty} |\alpha_n|^{1/n}}$. Hence, since $n\left|\frac{\alpha_n}{n!}\right|^{1/n} \sim e|\alpha_n|^{1/n}$, f is of exponential-type by Proposition 11.5 if and only if $\limsup_{n\to\infty} |\alpha_n|^{1/n} < \infty$, as was to be proved.

Definition. To the function f given by $f(z) = \Sigma \frac{\alpha_n}{n!} z^n$ we associate the function $\Phi(w) = \Sigma \alpha_n \frac{1}{w^{n+1}}$. Then f is of exponential-type if and only if Φ is holomorphic at ∞ and Φ is called the *Borel transform of f*.

Indeed, it is easily seen that we have a one-one linear correspondence between entire functions of exponential-type and functions Φ that are holomorphic near ∞ with $\Phi(\infty) = 0$.

Proposition 11.7.   *The Borel transform of f is an analytic continuation of the Laplace transform of f. More specifically, $\Phi(w) = \int_0^{+\infty} f(t)e^{-tw}\, dt$ in some right half-plane.*

Proof. First, $\int_0^{+\infty} \frac{t^n}{n!} e^{-tw}\, dt = \frac{1}{w^{n+1}}$ if $Rew > 0$. Now we estimate $\int_0^{+\infty}[f(t) - s_n(t)]e^{-tw}\, dt$, where $s_n(t)$ is the nth partial sum of $\Sigma \frac{\alpha_n}{n!} t^n$. We have, for $a_k = \frac{\alpha_k}{k!}$,

$$|f(z) - s_n(z)| = \sum_{n+1}^{\infty} |a_k|r^k = \sum_{n+1}^{\infty} |a_k|R^k \left(\frac{r}{R}\right)^k \leq M(R) \sum_{n+1}^{\infty} \left(\frac{r}{R}\right)^k$$

$$= \left(\frac{r}{R}\right)^{n+1} M(R)\frac{1}{1-\frac{r}{R}} \leq \left(\frac{r}{R}\right)^{n+1} e^{Rr'} \frac{R}{R-r},$$

where $\tau' > \tau$. Choose $R = 2r$ so that we get $|f(z) - s_n(z)| \leq \left(\frac{1}{2}\right)^n e^{2\tau'r}$, so that $|f(z) - s_n(z)| \leq e^{2\tau'|z|}$. Now $|f(t)| \leq e^{\tau't}$ and $|e^{-tw}| = e^{-tu}$, where $u = Rew$; thus we have convergence of the integral and we can interchange the summation and integration if we take $u > 2\tau'$. Thus we have

$$\int_0^{+\infty} f(t)e^{-tw}\, dt = \int_0^{+\infty} \left(\Sigma \left(\frac{\alpha_n}{n!}\right) t^n\right) e^{-tw}\, dt$$

$$= \Sigma \alpha_n \int_0^{+\infty} \frac{t^n}{n!} e^{-tw}\, dt = \Sigma \frac{\alpha_n}{w^{n+1}} = \Phi(w)$$

in some right half-plane, $u > 2\tau'$, as was to be proved.

As our final result in this chapter we shall give a direct proof that the order and type of f and f' are the same (Proposition 11.6).

Proof. Let $M_1(r) = \sup_\theta |f'(re^{i\theta})|$. Then we have from the Cauchy integral formula, $f'(z) = \frac{1}{2\pi i}\int_{|w|=R} \frac{f(w)dw}{(w-z)^2}$ if $R > r = |z|$, that $M_1(r) \leq M(r)\frac{R}{(R-r)^2}$. Now take $R = \lambda r$, where $\lambda > 1$. Then we get $M_1(r) \leq M(\lambda r)\frac{\lambda}{r(\lambda-1)^2} \leq M(\lambda r)\frac{\lambda}{(\lambda-1)^2}$ for $r > 1$. Hence, for $r > 1$,

$$\frac{\log^+ \log^+ M_1(r)}{\log r} \leq \frac{\log^+ \log^+ M(\lambda r)}{\log \lambda r}\frac{\log \lambda r}{\log r} + \frac{\log^+ \log^+ \frac{\lambda}{(\lambda-1)^2}}{\log r} + \frac{\log^+ 2}{\log r}$$

and thus $\rho_1 \leq \rho$.

In the other direction, supposing without loss of generality that $f(0) = 0$, we have $f(z) = \int_0^z f'(w)\, dw$, where we shall integrate along a ray passing through the origin. It follows that $M(r) \leq rM_1(r)$. Thus,

$$\frac{\log^+ \log^+ M(r)}{\log r} \leq \frac{\log^+ \log^+ r + \log^+ \log^+ M_1(r) + \log^+ 2}{\log r}$$

and hence $\rho \leq \rho_1$. Therefore $\rho = \rho_1$, as desired.

Similarly for type, we have

$$\frac{\log^+ M_1(r)}{r^\rho} \leq \frac{\log^+ M(\lambda r)}{(\lambda r)^\rho}\lambda^\rho + \frac{\log^+ \frac{\lambda}{(\lambda-1)^2}}{r^\rho}.$$

Hence $\tau_1 \leq \lambda^\rho \tau$ for any $\lambda > 1$ and thus $\tau_1 \leq \tau$. Also, $M(r) \leq rM_1(r)$, so that $\frac{\log^+ M(r)}{r^\rho} \leq \frac{\log^+ r}{r^\rho} + \frac{\log^+ M_1(r)}{r^\rho}$ and hence $\tau \leq \tau_1$. Therefore $\tau = \tau_1$.

12
Carleman's Theorem

Let f be holomorphic in Re $z \geq 0$ and suppose f has no zeros on $z = iy$. Choose $\rho > 0$ so that $\rho <$ (modulus of the smallest zero of f in Re $z \geq 0$). Let $\{z_n = r_n e^{i\theta_n}\}$ be the zeros of f in Re $z \geq 0$. Define the following:

$$\Sigma(R) = \Sigma(R:f) = \sum_{r_n \leq R} \left(\frac{1}{r_n} - \frac{r_n}{R^2} \right) \cos \theta_n$$

(proper multiplicity of the zeros taken into account);

$$I(R) = I(R:f) = \frac{1}{2\pi} \int_r^R \left(\frac{1}{t^2} - \frac{1}{R^2} \right) \log |f(it)f(-it)| \, dt,$$

where the integral is taken from ir to iR along the imaginary axis; and

$$J(R) = J(R:f) = \frac{1}{\pi R} \int_{-\pi/2}^{\pi/2} \log |f(Re^{i\theta})| \cos \theta \, d\theta,$$

where the integral is taken along the semicircle of radius R centered at 0. Then

$$\Sigma(R) = I(R) + J(R) + O(1).$$

Proof. Let Γ be the boundary of the "horseshoe" bounded by the semicircle of radius R, the semicircle of radius ρ, and the two vertical lines connecting them:

$$S = \frac{1}{2\pi i} \int_\Gamma \log f(z) \left[\frac{1}{z^2} + \frac{1}{R^2} \right] \, dz,$$

where we assume that f has no zeros on $|z| = R$ and that $\log f(z)$ denotes a branch of $\log f$ on Γ, i.e., $\log f(z)$ is some continuous function on Γ satisfying $\exp(\log f(z)) = f(z)$. The proof proceeds by evaluating the contour integral

$$(12.1) \qquad \int_{\substack{|z|=\rho \\ \text{Re } z \geq 0}} = O(1).$$

Along the negative imaginary axis $z = -iy$, $y > 0$, $dz = -i\,dy$, so

$$(12.2)$$
$$-\frac{1}{2\pi} \int_\rho^R \log f(-iy) \left[\frac{1}{R^2} - \frac{1}{y^2} \right] dy = \frac{1}{2\pi} \int_\rho^R \log f(-iy) \left[\frac{1}{y^2} - \frac{1}{R^2} \right] dy.$$

On $z = iy$, $y > 0$, $dz = i\,dy$, and we have

$$(12.3)$$
$$\frac{1}{2\pi} \int_R^\rho \log f(iy) \left[\frac{1}{(iy)^2} + \frac{1}{R^2} \right] dy = \frac{1}{2\pi} \int_\rho^R \log f(iy) \left[\frac{1}{y^2} - \frac{1}{R^2} \right] dy.$$

On $z = Re^{i\theta}$ (the large semicircle), $dz = iRe^{i\theta}\,d\theta$, so we have

$$(12.4)$$
$$\frac{1}{2\pi} \int_{-\pi/2}^{\pi/2} \log f(Re^{i\theta}) \frac{1}{R^2} \left(e^{-2i\theta} + 1 \right) e^{i\theta} R \, d\theta$$
$$= \frac{2}{2\pi R} \int_{-\pi/2}^{\pi/2} \log f(Re^{i\theta}) \cos\theta \, d\theta.$$

The sum of the real parts of (12.2) and (12.3) is

$$\frac{1}{2\pi} \int_\rho^R \log |f(iy)f(-iy)| \left(\frac{1}{y^2} - \frac{1}{R^2} \right) dy.$$

Thus

$$\text{Re } S = I(R) + J(R) + O(1).$$

Now integrate S by parts:

$$u = \log f(z) \qquad dv = \left(\frac{1}{z^2} + \frac{1}{R^2} \right) dz$$

$$du = \frac{f'(z)}{f(z)} dz \qquad v = \frac{z}{R^2} - \frac{1}{z}.$$

Hence,

$$S = \frac{1}{2\pi i} \left\{ \log f(z) \left[\frac{z}{R^2} - \frac{1}{z} \right] \right\} \Big|_{\text{start}}^{\text{finish}} - \frac{1}{2\pi i} \int_\Gamma \left(\frac{z}{R^2} - \frac{1}{z} \right) \frac{f'(z)}{f(z)} dz$$

$$= \text{purely imaginary} - \frac{1}{2\pi i} \int_\Gamma \left(\frac{z}{R^2} - \frac{1}{z} \right) \frac{f'(z)}{f(z)} dz.$$

On taking real parts and evaluating the remaining integral by the theory of residues we find

$$\text{Re } S = \sum_{|r_n| < R} \left(\frac{1}{r_n} - \frac{r_n}{R^2} \right) \cos \theta_n.$$

Remark 1. There is an obvious extension to meromorphic functions.

Remark 2. A formula of Nevanlinna (which stands in the same relation to Carleman's Theorem as the Poisson-Jensen formula to Jensen's formula) gives the value of f inside a semicircle from its values on the boundary and its zeros and poles. (See [5, p. 2].)

Next we present two applications of Carleman's Theorem.

Carlson's Theorem. *Suppose f is entire and of exponential-type $< \pi$ (i.e., $|f(z)| \le Ae^{B|z|}$, $B < \pi$) and $f(n) = 0$ for $n = 1, 2, \ldots$. Then $f \equiv 0$.*

Remark. Carlson's Theorem also is called the "sampling" theorem. Suppose $\varphi(t)$ is a continuous signal function with support in the interval $[-B, B]$ with $B < \pi$. Then the Fourier Transform of φ is

$$f(z) = \hat{\varphi}(z) = \frac{1}{\sqrt{2\pi}} \int_{\mathbb{R}} \varphi(t) e^{-izt} \, dt = \frac{1}{\sqrt{2\pi}} \int_{-B}^{B} \varphi(t) e^{-izt} \, dt.$$

Then, differentiation under the integral and a simple estimate shows that f is an entire function of exponential-type $B < \pi$. If f vanishes on the positive integers, then f vanishes everywhere by Carlson's Theorem and therefore $\varphi \equiv 0$ as well. By linearity, then, if we know f on the positive integers, we know it everywhere.

Proof. Suppose f does not vanish identically and satisfies the hypotheses of Carlson's Theorem. Then we may assume that f has no zeros on the imaginary axis. Otherwise, translate the plane $z \mapsto z - \epsilon$ for an appropriate $\epsilon > 0$. Recalling the notation in Carleman's Theorem, we observe that

$$\Sigma(r) \ge \sum_{|r_n| \le R} \left(\frac{1}{n} - \frac{n}{R^2} \right) \underset{\text{eff}}{\sim} \log R - O(1)$$

$$I(R) \le \frac{1}{\pi} \int_{\rho}^{R} \left(\frac{1}{t^2} - \frac{1}{R^2} \right) Bt \, dt \le \frac{1}{\pi} \int_{\rho}^{R} \frac{B}{t} \, dt \le \frac{B}{\pi} \log R$$

$$J(R) \le \frac{BR}{\pi R} 2 = \frac{2B}{\pi} = O(1).$$

Now applying Carleman's Theorem we see

$$\frac{B}{\pi} \log R + O(1) \le \log R.$$

But $\frac{B}{\pi} < 1$; eventually, this is a contradiction. f must vanish identically.

In Chapter 22 we will present a theorem of Malliavin-Rubel that gives a very sharp form of Carlson's Theorem.

For the next application, we prove a result that has applications to polynomial approximation theory.

Theorem. *Let α be the difference of two monotone functions on $[a, b]$ and let $f(z) = \int_a^b e^{zt}\, d\alpha(t)$. Suppose that $f(\lambda_n) = 0$ for each λ_n in a sequence Λ of numbers in the right half-plane satisfying $\Sigma Re\left(\frac{1}{\lambda_n}\right) = \infty$. Then $f(z) = 0$ for all z.*

Remark. By use of the Hahn-Banach and Stone-Weierstrass Theorems, it can be shown that finite sums of the form $\Sigma a_n \exp \lambda_n t$, $\lambda_n \in \Lambda$, are dense in the space of all continuous functions on $[a, b]$ in the uniform topology if and only if there is an α (actually, we must allow complex α, but there is no significant change) for which $\int \exp(\lambda_n t)\, d\alpha(t) = 0$ for each $\lambda_n \in \Lambda$ implies that $\int \exp(zt)\, d\alpha(t) = 0$ for all z. Our theorem above then will imply that if $\Sigma r_n^{-1} \cos \theta_n = \infty$, then the sums $\Sigma a_n \exp(\lambda_n t)$ are dense.

Proof. It is easily verified that f is an entire function. Unless f vanishes identically, we may suppose that f has no zeros on the imaginary axis, since we could otherwise translate the y axis. Now f is bounded on the imaginary axis, say $\log |f| \leq M$ there. Hence

$$I(R) \leq \frac{2M}{2\pi} \int_\rho^R \left(\frac{1}{t^2} - \frac{1}{R^2}\right)\, dt = O(1).$$

Now, $|f(z)| \leq A \max\{|e^{zt}| : a \leq t \leq b\}$, where A is a constant. If $c \geq \max(|a|, |b|)$, then $|f(z)| \leq Ae^{cz}$ so that $J(R) \leq O(1)$ also. Hence $\Sigma(R) = O(1)$.

We have

$$\Sigma(R) \geq \sum_{r_n \leq \frac{R}{2}} \left(\frac{1}{r_n} - \frac{r_n}{R^2}\right) \cos \theta_n \geq \frac{3}{4} \sum_{r_n \leq \frac{R}{2}} \left(\frac{1}{r_n} \cos \theta_n\right)$$

because $\frac{1}{r_n} - \frac{r_n}{R^2} \geq \frac{3}{4}\frac{1}{r_n}$ if $r_n \leq \frac{R}{2}$.

Thus

$$\sum_{r_n \leq \frac{R}{2}} \frac{1}{r_n} \cos \theta_n < \infty, \quad \text{and on letting} \quad R \to \infty;$$

we get a contradiction.

13
A Fourier Series Method

The idea presented in this chapter is the following: If f is a meromorphic function in the complex plane, and if

$$c_k(r, f) = \frac{1}{2\pi} \int_{-\pi}^{\pi} (\log |f(re^{i\theta})|) e^{-ik\theta} \, d\theta$$

is the kth Fourier coefficient of $\log |f(re^{i\theta})|$, then the behavior of $f(z)$ is reflected in the behavior of the sequence $\{c_k(r, f)\}$, and vice versa.

We prove a basic result in Theorem 13.4.5, which characterizes the rate of growth of f in terms of the rate of growth of the $c_k(r, f)$ and the density of the poles of f, generalizing Theorem 1 of [35]. We apply this theorem as in [35] to obtain estimates for some integrals involving $|f(z)|$ and to obtain information about the distribution of the zeros of an entire function from information about its rate of growth. Our presentation follows [36].

By these means, we make a study of certain general classes of meromorphic and entire functions that include many of the classically studied classes as special cases. Let $\lambda(r)$ be a positive, continuous, increasing, and unbounded function defined for all positive r. We say that the meromorphic function f is of finite λ-type to mean that there exist positive constants A and B with $T(r, f) \leq A\lambda(Br)$ for $r > 0$, where T is the Nevanlinna characteristic. An entire function f will be of finite λ-type if and only if there exist positive constants A and B such that

$$|f(z)| \leq \exp(A\lambda(B|z|)) \quad \text{for all complex} \quad z.$$

If we choose $\lambda(r) = r^\rho$, then the functions of finite λ-type are precisely the functions of growth not exceeding order ρ, finite exponential-type. We obtain here complete answers to certain basic questions about functions of finite λ-type. For example, in Theorem 13.5.2 we characterize the zero sets of entire functions of finite λ-type. This generalizes the well-known theorem of Lindelöf that corresponds to the classical case $\lambda(r) = r^\rho$. We obtain in Theorem 13.5.3 a corresponding result for meromorphic functions. Then, in Theorem 13.5.4, we give necessary and sufficient conditions on λ that each meromorphic function of finite λ-type be the quotient of two entire function of finite λ-type. In Chapter 14, we give Miles' proof that these conditions always hold.

The body of the chapter is divided into five sections, the last two of which contain the main results. The first three sections are concerned with various elementary, although sometimes complicated, results on sequences of complex numbers. The first section discusses the distribution of these sequences. The "Fourier coefficients" associated with a sequence are defined in the second section, and several technical propositions involving these coefficients also are proved there. The third section is concerned with the property of regularity of the function λ, which is closely connected with the algebraic structure of the field of meromorphic functions of finite λ-type. The fourth section contains the generalizations of the results of [35]. Finally, in the fifth section, the results about the distribution of zeros are proved.

We urge that, on a first reading, the reader read §4 first and then §5, referring to §1, §2, §3 for the appropriate definitions and statements of necessary preliminary results. After this, the complex sequence theory of the first three sections will seem much more natural.

13.1. An Analysis of Sequences of Complex Numbers

We study here the distribution of sequences $Z = \{z_n\}$, $n = 1, 2, 3, \ldots$, with multiplicity taken into account, of nonzero complex numbers z_n such that $z_n \to \infty$ as $n \to \infty$. Such sequences Z are studied in relation to so-called growth functions λ. We denote by A and B generic positive constants. The actual constants so represented may vary from one occurrence to the next. In many of the results, there is an implicit uniformity in the dependence of the constants in the conclusion on the constants in the hypotheses. For a more detailed explanation of this uniformity, we refer the reader to the remark following Proposition 13.1.11.

Let $Z = \{Z_n\}$ be a sequence of nonzero complex numbers such that $\lim z_n = \infty$ as $n \to \infty$.

Definition 13.1.1.   The counting function of Z is the function

$$n(r, Z) = \sum_{|z_n| \leq r} 1.$$

Definition 13.1.2. We define

$$N(r, Z) = \int_0^r \frac{n(t, Z)}{t} \, dt.$$

Proposition 13.1.3. *We have*

$$N(r, Z) = \sum_{|z_n| \le r} \log \frac{r}{|z_n|}.$$

Proof. Note that

$$\sum_{|z_n| \le r} \log \frac{r}{|z_n|} = \int_0^r \log \left(\frac{r}{t}\right) \, d[n(t, Z)].$$

The proposition follows from an integration by parts.

Proposition 13.1.4. *We have*

$$n(r, Z) = r \, \frac{d}{dr} N(r, Z).$$

Proof. Trivial.

Definition 13.1.5. We define, for $k = 1, 2, 3, \ldots$ and $r \ge 0$,

$$S(r; k : Z) = \frac{1}{k} \sum_{|z_n| \le r} \left(\frac{1}{z_n}\right)^k.$$

Definition 13.1.6. We define, for $k = 1, 2, 3, \ldots$ and $r_1, r_2 \ge 0$,

$$S(r_1, r_2; k : Z) = S(r_2; k : Z) - S(r_1; k : Z).$$

When no confusion will result, we will drop the Z from the above notation and write $n(r)$, $S(r; k)$, etc.

Definition 13.1.7. A *growth function* $\lambda(r)$ is a function defined for $0 < r < \infty$ that is positive, nondecreasing, continuous, and unbounded.

Throughout this chapter, λ will always denote a growth function.

Definition 13.1.8. We say that the sequence Z has *finite λ-density* to mean that there exist constants A, B such that, for all $r > 0$,

$$N(r, Z) \le A\lambda(Br).$$

Proposition 13.1.9. *If Z has finite λ-density, then there are constants A, B such that*

$$n(r, Z) \le A\lambda(Br).$$

Proof. We have

$$n(r, Z) \log 2 \le \int_r^{2r} \frac{n(t, Z)}{t} \, dt \le N(2r, Z).$$

Definition 13.1.10. We say that the sequence Z is λ-*balanced* to mean that there exist constants A, B such that

$$(13.1.1) \qquad |S(r_1, r_2; k : Z)| \le \frac{A\lambda(Br_1)}{r_1^k} + \frac{A\lambda(Br_2)}{r_2^k}$$

for all r_1, $r_2 > 0$ and $k = 1, 2, 3, \ldots$. We say that Z is *strongly λ-balanced* to mean that

$$(13.1.2) \qquad |S(r_1, r_2; k : Z)| \le \frac{A\lambda(Br_1)}{kr_1^k} + \frac{A\lambda(Br_2)}{kr_2^k}$$

for all r_1, $r_2 > 0$ and $k = 1, 2, 3, \ldots$.

Proposition 13.1.11. *If Z has finite λ-density and is λ-balanced, then Z is strongly λ-balanced.*

Remark. Using this result for illustrative purposes, we make explicit here the uniformity that we leave implicit in the statements of similar results. The assertion is that if Z has finite λ-density with implied constants A, B, and is λ-balanced with implied constants A', B', then Z is strongly λ-balanced with implied constants A'', B'' that depend only on A, B, A', B' and not on Z or λ.

Proof of Proposition 13.1.11. We observe first that, if $r > 0$, and if we let $r' = rk^{1/k}$, then

$$(13.1.3) \qquad |S(r, r'; k)| \le \frac{3n(r')}{kr^k}.$$

To prove this we note that

$$|S(r, r'; k)| \le \frac{1}{k} \int_r^{r'} \frac{1}{t^k} dn(t),$$

from which (13.1.3) follows after an integration by parts. Now, for r_1, $r_2 > 0$, let $r_1' = r_1 k^{1/k}$ and $r_2' = r_2 k^{1/k}$. Then

$$|S(r_1, r_2; k)| \le |S(r_1', r_2'; k)| + |S(r_1, r_1'; k)| + |S(r_2, r_2'; k)|.$$

On combining this inequality with (13.1.3), Proposition 13.1.9, and the fact that $k^{1/k} \leq 2$, we have

$$|S(r_1, r_2; k)| \leq |S(r_1', r_2'; k)| + \frac{1}{kr_1^k} A\lambda(Br_1) + \frac{1}{kr_2^k} A\lambda(Br_2).$$

But, by hypothesis,

$$|S(r_1', r_2'; k)| \leq \frac{1}{kr_1^k} A\lambda(Br_1) + \frac{1}{kr_2^k} A\lambda(Br_2)$$

for $k = 1, 2, 3, \ldots$.

Definition 13.1.12.   We say that the sequence Z is λ-*poised* to mean that there exists a sequence α of complex numbers $\alpha = \{\alpha_k\}$, $k = 1, 2, 3, \ldots$ such that, for some constants A, B, we have, for $k = 1, 2, 3, \ldots$ and $r > 0$,

$$(13.1.4) \qquad |\alpha_k + S(r; k : Z)| \leq \frac{A\lambda(Br)}{r^k}.$$

If the following stronger inequality:

$$(13.1.5) \qquad |\alpha_k + S(r; k : Z)| \leq \frac{A\lambda(Br)}{kr^k}$$

holds, we say that Z is *strongly λ-poised*.

Proposition 13.1.13.   *If Z has finite λ-density and is λ-poised, then Z is strongly λ-poised.*

Proof. The proof is quite analogous to the proof of Proposition 13.1.11, based on the substitution $r' = rk^{1/k}$. We omit the details.

Proposition 13.1.14.   *A sequence Z is λ-balanced if and only if it is λ-poised, and is strongly λ-balanced if and only if it is strongly λ-poised.*

Proof. We prove only the second assertion, since the proof of the first assertion is virtually the same. If it is first supposed that Z is strongly λ-poised, where $\{\alpha_k\}$ is the relevant sequence, then we have

$$|S(r_1, r_2; k)| = |S(r_2; k) + \alpha_k - \alpha_k - S(r_1; k)|$$
$$\leq |\alpha_k + S(r_1; k)| + |\alpha_k + S(r_2; k)|,$$

so that Z is strongly λ-balanced.

Suppose now that Z is strongly λ-balanced, with A, B being the relevant constants. Let

$$p(\lambda) = \inf\{p = 1, 2, 3, \cdots : \liminf_{r \to \infty} \frac{\lambda(r)}{r^p} = 0\}.$$

Naturally, we let $p(\lambda) = \infty$ in the case $\liminf \lambda(r)r^p > 0$ as $r \to \infty$ for each positive integer p. For $1 \le k < p(\lambda)$, we have $\inf r^{-k}\lambda(Br) > 0$ for $r > 0$. Thus, there exist positive numbers r_k such that

$$\frac{\lambda(Br_k)}{r_k^k} < 2\frac{\lambda(Br)}{r^k}$$

for $r > 0$ and $1 \le k < p(\lambda)$. For k in this range, we define

(13.1.6) $\alpha_k = -S(r_k; k).$

For those k, if there are any, for which $k \ge p(\lambda)$, we choose a sequence $0 < \rho_1 < \rho_2 < \cdots$ with $\rho_j \to \infty$ as $j \to \infty$ such that

$$\lim_{j \to \infty} \frac{\lambda(B\rho_j)}{\rho_j^{p(\lambda)}} = 0.$$

For values of k, then, such that $k \ge p(\lambda)$, we define

(13.1.7) $\alpha_k = \lim_{j \to \infty} -S(\rho_j; k).$

To show that the limit exists, we prove that the sequence $\{S(\rho_j; k)\}$, $j = 1, 2, \ldots$, is a Cauchy sequence. Let

$$\Delta_{j,m} = S(\rho_j; k) - S(\rho_m; k) = S(\rho_m, \rho_j; k).$$

We have

$$|\Delta_{j,m}| \le \frac{A\lambda(B\rho_m)}{k\rho_m^k} + \frac{A\lambda(B\rho_j)}{k\rho_j^k}.$$

Since $\rho^k \ge \rho^{p(\lambda)}$ for $\rho \ge 1$, it follows from the choice of the ρ_j that $\Delta_{j,m} \to 0$ as $j, m \to \infty$. We now claim that

$$|\alpha_k + S(r; k)| \le \frac{3A\lambda(Br)}{kr^k}.$$

For, if $1 \le k < p(\lambda)$, then

$$|\alpha_k + S(r; k)| = |S(r_k, r; k)| \le \frac{A\lambda(Br)}{kr^k} + \frac{A\lambda(Br_k)}{kr^k} \le \frac{3A\lambda(Br)}{kr^k};$$

if $k \ge p(\lambda)$, then

$$|\alpha_k + S(r; k)| = \lim_{j \to \infty} |S(r, \rho_j; k)| \le \frac{A\lambda(Br)}{kr^k} + \limsup_{j \to \infty} \frac{A\lambda(B\rho_j)}{k\rho_j^k} = \frac{A\lambda(B\rho)}{k\rho^k}.$$

Definition 13.1.15. We say that the sequence Z is λ-*admissible* to mean that Z has *finite* λ-*density* and is λ-*balanced*.

In view of Propositions 13.1.11 and 13.1.13, the following result is immediate.

Proposition 13.1.16. *Suppose that Z has finite λ-density. Then the following are equivalent:*

(i) Z is λ-balanced;
(ii) Z is strongly λ-balanced;
(iii) Z is λ-poised;
(iv) Z is strongly λ-poised;
(v) Z is λ-admissible.

In Proposition 13.3.3, we give a simple characterization of λ-admissible sequences in the special case $\lambda(r) = r^\rho$.

13.2. The Fourier Coefficients Associated with a Sequence

We now present the sequence of so-called Fourier coefficients associated with a sequence Z of complex numbers, and study its properties. We will use it in §5 to construct an entire function f whose zero set coincides with Z, and to determine some properties of entire and meromorphic functions whose growth is restricted. The reason for calling them "Fourier coefficients" will become apparent on comparing their definition with Lemma 13.4.2.

Definition 13.2.1. We define, for $k = 1, 2, 3, \ldots$,

$$S'(r; k : Z) = \frac{1}{k} \sum_{|z_n| \leq r} \left(\frac{\bar{z}_n}{r} \right)^k.$$

Proposition 13.2.2. *We have*

$$|S'(r; k : Z)| \leq \frac{1}{k} N(er, Z).$$

Proof. It is clear that $|S'(r; k : Z)| \leq n(r)/k$, and we also have

$$n(r) \leq \int_r^{er} \frac{n(t)}{t} \, dt \leq N(er).$$

Definition 13.2.3. Let $\alpha = \{\alpha_k\}$, $k = 1, 2, 3, \ldots$, be a sequence of complex numbers. The sequence $\{c_k(r; Z : \alpha)\}$, $k = 0, \pm 1, \pm 2, \ldots$, defined by

$$(13.2.1) \qquad c_0(r; Z : \alpha) = c_0(r; Z) = N(r, Z),$$

$$(13.2.2) \qquad \begin{aligned} c_k(r; Z : \alpha) =& \frac{r^k}{2}\{\alpha_k + S(r; k : Z)\} \\ & - \frac{1}{2}S'(r; k : Z) \quad \text{for} \quad k = 1, 2, 3 \ldots, \end{aligned}$$

$$(13.2.3) \qquad c_{-k}(r; Z : \alpha) = \overline{(c_k(r; Z; \alpha))} \quad \text{for} \quad k = 1, 2, 3 \ldots,$$

is said to be a sequence of *Fourier coefficients associated with* Z.

Definition 13.2.4. A sequence $\{c_k(r; Z : \alpha)\}$ of Fourier coefficients associated with Z is called λ-*admissible* if there exist constants A, B such that

$$(13.2.4) \qquad |c_k(r : z; \alpha)| \leq \frac{A\lambda(Br)}{|k| + 1} \quad (k = 0, \pm 1, \pm 2, \ldots).$$

Proposition 13.2.5. *A sequence Z is λ-admissible if and only if there exists a λ-admissible sequence of Fourier coefficients associated with Z.*

Proof. Suppose that Z is λ-admissible. Then, by Proposition 13.1.16, Z is strongly λ-poised. Let $\alpha = (\alpha_k)$, $k = 1, 2, 3, \ldots$, be the relevant constants, and form $\{c_k(r; Z : \alpha)\}$ from them by means of (13.2.1)–(13.2.3). Now Definition 13.2.4 holds for $k = 0$ and some constants A, B since Z has finite λ-density. For $k = \pm 1, \pm 2, \pm 3, \ldots$, we have

$$|c_k(r; Z : \alpha)| \leq \frac{r^{|k|}}{2}|\alpha_k + S(r; k)| + \frac{1}{2}|S'(r; k)|.$$

Then an inequality of the form (13.2.4) holds by Proposition 13.2.2 since Z has finite λ-density, and because Z is strongly λ-poised with respect to the constants $\{\alpha_k\}$.

On the other hand, suppose that (13.2.4) holds. Then

$$N(r) = c_0(r) \leq A\lambda(Br),$$

so that Z has finite λ-density. Moreover,

$$\begin{aligned} \left|\frac{r^k}{2}(\alpha_k + S(r; k))\right| &= \left|c_k(r; Z : \alpha) + \frac{1}{2}S'(r; k)\right| \\ &\leq \frac{A\lambda(Br)}{|k| + 1} + \frac{N(er)}{2k} \leq \frac{2A\lambda(eBr)}{k}, \end{aligned}$$

so that Z is strongly λ-poised. By Proposition 13.1.16, it follows that Z is λ-admissible.

Proposition 13.2.6. *Suppose that Z and $\alpha = \{\alpha_k\}$ are such that $|c_k(r; Z : \alpha)| \leq A\lambda(Br)$. Then $\{c_k(r; Z : \alpha)\}$ is λ-admissible. In particular, there exist constants A', B', depending only on A, B, such that*

$$|c_k(r; Z : \alpha)| \leq \frac{A'\lambda(B'r)}{|k| + 1|}.$$

Proof. For $k = 1, 2, \ldots$, we have

$$(13.2.5) \qquad |c_k(r)| \leq \frac{r^k}{2}|\alpha_k + S(r; k)| + \frac{1}{2}|S'(r; k)|$$

and

$$(13.2.6) \qquad \frac{r^k}{2}|\alpha_k + S(r; k)| \leq |c_k(r)| + \frac{1}{2}|S'(r; k)|.$$

Since $c_0(r) = N(r) \leq A\lambda(Br)$, Z has finite λ-density. Then, by Proposition (13.2.2), $|S'(r; k)| \leq (1/k)O(\lambda(O(r)))$ uniformly for $k = 1, 2, 3, \ldots$, by which we mean that there are constants A'', B'' for which $|S'(r, k)| \leq (1/k)A''\lambda(B''r)$. From our hypothesis and (13.2.6), it then follows that

$$r^k|\alpha_k + S(r; k)| = O(\lambda(O(r))) \quad \text{uniformly for} \quad k > 0.$$

Then, by Proposition 13.1.13, we have that

$$r^k|\alpha_k + S(r; k)| \leq \frac{1}{k}O(\lambda(O(r))) \quad \text{uniformly for} \quad k = 1, 2, 3 \ldots.$$

Then, using (13.2.5), we have

$$|c_k(r)| \leq \frac{1}{k}O(\lambda(O(r))) \quad \text{uniformly for} \quad k = 1, 2, 3 \ldots.$$

Since $c_{-k}(r) = \overline{(c_k(r))}$, and since Z has finite upper λ-density, the proposition follows immediately.

Definition 13.2.7. The *quadratic semi-norm of a sequence* $\{c_k(r; Z : \alpha)\}$ of Fourier coefficients associated with Z is given by

$$E_2(r; Z : \alpha) = \left\{ \sum_{k=-\infty}^{\infty} |c_k(r; Z : \alpha)|^2 \right\}^{1/2}.$$

Proposition 13.2.8. *The Fourier coefficients $\{c_k(r; Z : \alpha)\}$ are λ-admissible if and only if $E_2(r; Z : \alpha) \leq A\lambda(Br)$ for some constants A, B.*

Proof. First, if

$$|c_k(r; Z : \alpha)| \leq \frac{A_1 \lambda(B_1 r)}{|k| + 1},$$

then $E_2(r; Z : \alpha) \leq A\lambda(Br)$, where $B = B_1$ and

$$A = A_1 \left\{ \sum_{k=-\infty}^{\infty} \left(\frac{1}{|k| + 1} \right)^2 \right\}^{1/2}.$$

On the other hand, suppose there are constants A, B for which $E_2(r; Z : \alpha) \leq A\lambda(Br)$. Then it is clear that $|c_k(r; Z : \alpha)| \leq A\lambda(Br)$, so that by Proposition 13.2.6, $\{c_k(r; Z : \alpha)\}$ is λ-admissible.

13.3. Sequences That Are λ-Balanceable

In this section, we are concerned with the process of enlarging a sequence Z so that it becomes λ-balanced. Growth functions λ for which this is always possible are called *regular* and give rise to associated fields of meromorphic functions with special properties; for example, see Theorem 13.5.4. The principal results of this section are Propositions 13.3.5 and 13.3.6, which give the simple condition that λ be regular. In addition, we give in Proposition 13.3.3 a simple characterization of λ-admissible sequences of the case $\lambda(r) = r^\rho$.

Definition 13.3.1. The sequence Z is λ-*balanceable* if there exists a λ-admissible supersequence Z' of Z.

Definition 13.3.2. The growth function λ is *regular* if every sequence Z that has finite λ-density is λ-balanceable.

Proposition 13.3.3. *Suppose that $\lambda(r) = r^\rho$, where $\rho > 0$. Then*
(i) the sequence Z is of finite λ-density if and only if $\limsup r^{-\rho} n(r, Z) < \infty$ as $r \to \infty$;
(ii) if ρ is not an integer, then every sequence of finite λ-density is λ-admissible;
(iii) if ρ is an integer, then Z is λ-admissible if and only if Z is of finite λ-density and $S(r; \rho : Z)$ is a bounded function of r;
(iv) the function $\lambda(r) = r^\rho$ is regular.

Proof. To prove (i), we have that $n(r) = O(r^\rho)$ whenever Z has finite λ-density. On the other hand, if $\limsup r^{-\rho} n(r) < \infty$, then $n(r) \leq Ar^\rho$ for some positive constant A, so that

$$N(r) = \int_0^r t^{-1} n(t) \, dt \leq A\rho^{-1} r^\rho.$$

To prove (ii), suppose that $N(t) \leq At^\rho$. Then so long as $k \neq \rho$, we have

$$(13.3.1) \qquad \int_{r_1}^{r_2} \frac{1}{t^k} \, dn(t) \leq \left(A + \frac{A}{|\rho - k|} \right) \left(\frac{\lambda(r_1)}{r_1^k} + \frac{\lambda(r_2)}{r_2^k} \right).$$

For, on integrating by parts, we have that the integral is equal to

$$\frac{n(r_2)}{r_2^k} - \frac{n(r_1)}{r_1^k} + k \int_{r_1}^{r_2} \frac{n(t)}{t^{k+1}} \, dt.$$

But

$$\frac{n(r_2)}{r_2^k} \leq A \frac{r_2^\rho}{r_2^k} = \frac{A\lambda(r_2)}{r_2^k},$$

and similarly

$$\frac{n(r_1)}{r_1^k} \leq \frac{A\lambda(r_1)}{r_1^k}.$$

Moreover,

$$\int_{r_1}^{r_2} \frac{n(t)}{t^{k+1}} \, dt \leq A \int_{r_1}^{r_2} \frac{t^\rho}{t^{k+1}} \, dt \leq \frac{A}{|\rho - k|} \left(\frac{r_2^\rho}{r_2^k} + \frac{r_1^\rho}{r_1^k} \right),$$

and the inequality (13.3.1) follows. Hence, so long as ρ is not an integer, every sequence Z of finite r^ρ-density is r^ρ-balanced.

To prove (iii), suppose that Z has finite r^ρ-density and that ρ is an integer. Then, by (13.3.1), we see that all the conditions that Z be λ-balanced are satisfied except for $k = \rho$. For this case, the condition that $S(r_1, r_2; \rho)$ be bounded by $r_1^{-\rho} A\lambda(Br_1) + r_2^{-\rho} A\lambda(Br_2)$ for some A, B is precisely the condition that $S(r; \rho)$ be bounded, as is quite easy to see.

To prove (iv), we observe first that if ρ is not an integer, then $\lambda(r) = r^\rho$ is trivially regular by (ii). If ρ is an integer and Z has finite r^ρ-density, let Z' be the sequence obtained by adding to Z all numbers of the form $\omega^{-1}Z$, where $\omega^\rho = 1$, but $\omega \neq 1$. Then Z' has finite r^ρ-density and $S(r; \rho : Z') = 0$ for all $r > 0$. Hence, by (iii), Z' is r^ρ-admissible, and it follows that $\lambda(r) = r^\rho$ is regular.

The next two results give simple conditions, both satisfied in case $\lambda(r) = r^\rho$, that imply that λ is regular.

Definition 13.3.4. We say that the growth function λ is *slowly increasing* to mean that $\lambda(2r) \leq M\lambda(r)$ for some constant M.

If λ is slowly increasing, it is easy to show that for some positive number ρ, $\lambda(r) = O(r^\rho)$ as $r \to \infty$.

Proposition 13.3.5. *If λ is slowly increasing, then λ is regular.*

Proposition 13.3.6. *If $\log \lambda(e^x)$ is convex, then λ is regular.*

The proofs of these results use the next lemma.

Lemma 13.3.7. *The growth function λ is slowly increasing if and only if there exist an integer p_0 and constants A, B such that*

$$(13.3.2) \qquad \int_r^\infty \frac{\lambda(t)}{t^{p+1}}\, dt \le \frac{A\lambda(Br)}{pr^p} \quad \text{for} \quad p \ge p_0 > 0.$$

If $\lambda(r) = r^\rho$, then we may choose $p_0 = [\rho] + 1$.

Proof. Suppose first that (13.3.2) holds. We may clearly suppose that $B \ge 1$. Then

$$\frac{A\lambda(Br)}{pr^p} \ge \int_r^\infty \frac{\lambda(t)}{t^{p+1}}\, dt \ge \int_{2Br}^\infty \frac{\lambda(t)}{t^{p+1}}\, dt \ge \frac{\lambda(2Br)}{p(2B)^p r^p}$$

whenever $p \ge p_0$. Taking $p = p_0$, we have $\lambda(2Br) \ge M\lambda(Br)$, where

$$M = A(2B)^{p_0},$$

so that $\lambda(r)$ is slowly increasing. Suppose next that $\lambda(r)$ is slowly increasing, say $\lambda(2r) \le M\lambda(r)$. Then

$$\int_r^\infty \frac{\lambda(t)}{t^{p+1}}\, dt = \sum_{k=0}^\infty \int_{2^k r}^{2^{k+1}r} \frac{\lambda(t)}{t^{p+1}}\, dt \le \sum_{k=0}^\infty \frac{\lambda(2^{k+1}r)}{pr^p(2^k)^p} \le M\frac{\lambda(r)}{pr^p} \sum_{k=0}^\infty \left(\frac{M}{2^p}\right)^k.$$

Hence, if p_0 is taken so large that $2^{p_0} > M$, we have an inequality of the form (13.3.2). In case $\lambda(r) = r^\rho$, we have $M = 2^\rho$, and the final assertion follows.

Proof of Proposition 13.3.5. Let λ be slowly increasing, and let Z be a sequence of finite λ-density. Choose p_0 as in the last lemma so that, for $p \ge p_0$,

$$\int_r^\infty \frac{\lambda(t)}{t^{p+1}}\, dt \le \frac{A\lambda(Br)}{pr^p}.$$

Define

$$Z' = \bigcup_{k=0}^{p_0} \omega^{-k} Z,$$

where $n_0 = p_0 + 1$, $\omega = \exp(2\pi i/n_0)$, and $w^{-k}Z = \{\omega^{-k} z_n\}$, $n = 1, 2, 3, \ldots$. Then we have $S(r_1, r_2; k : Z') = 0$ for $k = 1, 2, \ldots, p_0$, since

$$1 + \omega^k + \omega^{2k} + \cdots + \omega^{p_0 k} = 0$$

so long as $\omega^k \ne 1$, and this is true for $k = 1, 2, \ldots, p_0$. Hence, to prove that Z' is λ-balanced, we need consider only $k > p_0$. For such k, with $r < r'$, we have

$$|S(r, r'; k : Z)| = \frac{1}{k}\left| \sum_{r < |z_n| < r'} \left(\frac{1}{z_n}\right)^k \right| \le \frac{1}{k} \int_r^{r'} \frac{1}{t^k}\, dn(t, Z').$$

On integrating by parts, we have

$$\left| \int_r^{r'} \frac{1}{t^k}\, dn(t, Z') \right| \leq \frac{n(r', Z')}{(r')^k} + \frac{n(r, Z')}{r^k} + k \int_r^{r'} \frac{n(t, Z')}{t^{k+1}}\, dt.$$

Since Z is of finite λ-density and $n(r, Z') = (p_0 + 1)n(r, Z)$, we have $n(r, Z') \leq A_1 \lambda(B_1 r)$ for some constants A_1, B_1 by Proposition 13.1.9.

Since λ is slowly increasing, we have $\lambda(B_1 r) \leq A_2 \lambda(r)$ for some positive constant $A_2 > 0$. To complete the proof of the proposition, we have only to prove that

$$\int_r^{r'} \frac{n(t, Z')}{t^{k+1}}\, dt \leq \frac{A' \lambda(B' r)}{k r^k}$$

for some constants A', B'. However,

$$\int_r^{r'} \frac{n(t, Z')}{t^{k+1}}\, dt \leq A_2 \int_r^{\infty} \frac{\lambda(t)}{t^{k+1}} \leq \frac{A A_2 \lambda(Br)}{k r^k}$$

since $k > p_0$.

Proof of Proposition 13.3.6. It is no loss of generality to suppose that $r^{-p}\lambda(r) \to \infty$ as $r \to \infty$ for each $p > 0$, since otherwise λ is slowly increasing by Lemma 13.3.7, and then Proposition 13.3.5 applies. Now for $p = 1, 2, 3, \ldots$, let R_p be the largest number such that

$$\frac{\lambda(R_p)}{R_p^p} = \inf_{r > 0} \frac{\lambda(r)}{r^p},$$

and let $R_0 = 0$. Then we have that $R_0 \leq R_1 \leq R_2 \leq \ldots$, and that $R_p \to \infty$ as $p \to \infty$. Further, by Lemma 13.3.7, $r^{-p}\lambda(r)$ decreases for $r \leq R_p$ and increases for $r \geq R_p$. We also have the inequality

$$(13.3.3) \qquad 2^p \lambda(r) \leq 2\lambda(2r) \quad \text{if} \quad r \geq R_{p-1},$$

since, by the above remark,

$$\frac{\lambda(r)}{r^{p-1}} \leq \frac{\lambda(2r)}{(2r)^{p-1}}.$$

Now let Z be of finite λ-density. For convenience of notation, we suppose that $N(r) \leq \lambda(r)$ and $n(r) \leq \lambda(r)$, since we could otherwise replace the function $\lambda(r)$ by the function $A\lambda(Br)$ for suitable constants A, B. We then claim that

$$(13.3.4) \qquad \int_r^{r'} \frac{1}{t^k}\, dn(t) \leq \frac{4\lambda(r)}{r^k}$$

if $k \geq 2^p$ and $r \leq r' \leq R_p$.

To prove (13.3.4), we first integrate by parts, replacing the integral by

$$\frac{n(r')}{(r')^k} - \frac{n(r)}{r^k} + k \int_r^{r'} \frac{n(t)}{t^{k+1}} \, dt.$$

Now

$$\frac{n(r')}{(r')^k} \leq \frac{\lambda(r')}{(r')^k} \leq \frac{\lambda(r)}{r^k}$$

since $r \leq r'$ and $r^{-k}\lambda(r)$ is decreasing for $r \leq R_p \leq R_k$. Also,

$$\int_r^{r'} \frac{n(t)}{t^{k+1}} \, dt \leq \int_r^{r'} \frac{\lambda(r)}{r^p} \frac{1}{t^{k+1-p}} \, dt \leq \frac{\lambda(r)}{r^p} \frac{1}{(k-p)} \frac{1}{r^{k-p}},$$

since $t^{-p}\lambda(t) \leq r^{-p}\lambda(r)$ for $r \leq t \leq r' \leq R_p$. Thus,

$$\int_r^{r'} \frac{1}{t^k} \, dn(t) \leq \frac{\lambda(r)}{r^k} + \frac{\lambda(r)}{r^k} + \frac{k}{(k-p)} \frac{\lambda(r)}{r^k}.$$

We have $k/k(k-p) \leq 2$ since $k \geq 2^p$, and (13.3.4) follows.

We now define Z' as follows. For each $z_n \in Z$ with $R_{p-1} < |z_n| \leq R_p$, we introduce into Z' the numbers

$$z_n, \quad \frac{1}{\omega} z_n, \ldots, \frac{1}{\omega^{m-1}} z_n,$$

where $m = m(p) = 2^p$ and $\omega = \omega(m) = \exp(2\pi i/m)$. We make the following assertions:

(13.3.5) $n(r, Z') - n(R_{p-1}, Z') = 2^p(n(r) - n(R_{p-1}))$ if $R_{p-1} \leq r \leq R_p$,

(13.3.6) $\qquad\qquad n(r, Z') \leq 2^p n(r)$ if $r \leq R_p$,

(13.3.7) $\qquad\qquad N(r, Z') \leq 2^p \lambda(r)$ if $r \leq R_p$,

(13.3.8)
$$\int_r^{r'} \frac{1}{t^k} \, dn(t, Z') \leq k \int_r^{r'} \frac{1}{t^k} \, dn(t) \quad \text{if} \quad k \geq 2^p \quad \text{and} \quad r \leq r' \leq R_p.$$

(13.3.9)
$\qquad S(r, r'; k : Z') = 0$ if $r, r' \geq R_p$ and k is not a multiple of 2^p.

The assertions (13.3.5) and (13.3.6) follow immediately from the definition of Z', while (13.3.7) follows from (13.3.6) and (13.3.8) follows easily from (13.3.5). To prove (13.3.9), it is enough to prove that $S(r, r'; k : Z') = 0$ if $R_{j-1} \leq r \leq r' \leq R_j$, $j \geq p$, and k is not a multiple of 2^p. But, in this case, we have

$$S(r, r'; k : Z') = \gamma S(r, r'; k : Z),$$

where

$$\gamma = 1 + \omega^k + \omega^{2k} + \cdots + \omega^{(m-1)k},$$

where $m = m(j) = 2^j$ and $\omega = \omega(m) = \exp(2\pi i/m)$. Since k is not a multiple of 2^p, k is therefore certainly not a multiple of 2^j, so that $\omega^k \neq 1$. We then have

$$\gamma = \frac{1 - \omega^{km}}{1 - \omega^k} = 0,$$

and our assertion is proved.

We now prove that Z' is λ-admissible. To see that Z' has finite λ-density, let $r > 0$ and let p be such that $R_{p-1} \leq r \leq R_p$. Then, by (13.3.7) and (13.3.3), we have that $N(r, Z') \leq 2^p \lambda(r) \leq 2\lambda(2r)$. To see that Z' is λ-balanced, let k be a positive integer and suppose that $0 < r \leq r'$. Write k in the form $2^p q$, where q is odd. Then, by (13.3.9), $S(r, r'; k : Z') = 0$ if $R_p \leq r < r'$. Suppose that $r \leq R_p$. Then $S(r, r'; k : Z') = S(r, r''; k : Z')$, where $r'' = \min(r', R_p)$, by (13.3.9). However,

$$|S(r, r''; k : Z')| \leq \frac{1}{k} \int_r^{r''} \frac{1}{t^k} \, dn(t, Z').$$

By (13.3.8), this last term does not exceed

$$\int_r^{r''} \frac{1}{t^k} \, dn(t),$$

and this, in turn, does not exceed $4r^{-k}\lambda(r)$, by (13.3.4). Consequently, we always have $|S(r, r'; k : Z)| \leq 4r^{-k}\lambda(r)$, so that Z' is λ-balanced, and the proof is complete.

13.4. The Fourier Coefficients Associated with a Meromorphic Function

In this section, we associate a Fourier series with a meromorphic function and use it to study properties of the function. As we mentioned at the beginning, the results of this section are generalized versions of the results of the earlier paper [35], and the proofs are essentially the same. Our notation follows the notation of [35] and the usual notation from the theory of meromorphic and entire functions. Our presentation still follows [36]. We first recall the results from the theory of meromorphic functions that will be needed.

For a nonconstant meromorphic function f, we denote by $Z(f)$ [respectively $W(f)$] the sequence of zeros (respectively poles) of f, each occurring the number of times indicated by its multiplicity. We suppose throughout that $f(0) \neq 0, \infty$. It requires only minor modifications to treat the case where $f(0) = 0$ or $f(0) = \infty$. By $n(r, f)$ we denote the number of poles of f in the disc $\{z : |z| \leq r\}$. By $N(r, f)$ we denote the function

$$N(r, f) = \int_0^r \frac{n(t, f)}{t} \, dt,$$

and by $m(r, f)$ the function

$$m(r, f) = \frac{1}{2\pi} \int_{-\pi}^{\pi} \log^+ |f(re^{i\theta})| \, d\theta,$$

where $\log^+ x = \max(\log x, 0)$. We have, of course, that $n(r, f) = n(r, W(f))$ and $N(r, f) = N(r, W(f))$. The Nevanlinna characteristic, which measures the growth of f, is the function

$$T(r, f) = m(r, f) + N(r, f).$$

Three fundamental facts about $T(r, f)$ are that

(13.4.1)
$$T(r, f) = T\left(r, \frac{1}{f}\right) + \log |f(0)|,$$

(13.4.2)
$$T(r, fg) \leq T(r, f) + T(r, g),$$

(13.4.3)
$$T(r, f + g) \leq T(r, f) + T(r, g) + \log 2.$$

An easy consequence of (13.4.1) is that

(13.4.4)
$$\frac{1}{2\pi} \int_{-\pi}^{\pi} \left| \log |f(re^{i\theta})| \right| \, d\theta \leq 2T(r, f) + \log |f(0)|.$$

This follows from (13.4.1) by observing that the first term is equal to $m(r, f) + m(r, 1/f)$, which is dominated by $T(r, f) + T(r, 1/f)$.

For the entire functions f, we use the notation

$$M(r, f) = \sup\{|f(z)| : z = r\}.$$

The following inequality relates these two measures of the growth of f in case f is entire:

(13.4.5)
$$T(r, f) \leq \log^+ M(r, f) \leq \frac{R + r}{R - r} T(R, f)$$

for $0 \leq r \leq R$. We will use (13.4.5) mostly in the form

(13.4.6)
$$T(r, f) \leq \log^+ M(r, f) \leq 3T(2r, f),$$

which results from setting $R = 2r$ in (13.4.5).

The following lemma, which is fundamental in our method, was proved in [9] and [35].[†] We reproduce the proof of [35] here.

[†] I have a vague memory of seeing this formula in a paper of Frithiof Nevanlinna published around 1925, but was unable to find it on a recent search.

Lemma 13.4.1. *If $f(z)$ is meromorphic in $|z| \leq R$, with $f(0) \neq 0$, ∞, and $Z(f) = \{z_n\}$, $W(f) = \{w_n\}$, and if $\log(f(z)) = \sum_{k=0}^{\infty} \alpha_k z^k$ near $z = 0$, then for $0 < r \leq R$ we have*

$$(13.4.7) \qquad \log|f(re^{i\theta})| = \sum_{k=-\infty}^{\infty} c_k(r, f) e^{ik\theta},$$

where the $c_k(r, f)$ are given by

$$(13.4.8) \qquad \begin{aligned} c_0(r, f) &= \log|f(0)| + \sum_{|z_n| \leq r} \log \frac{r}{|z_n|} - \sum_{|w_n| \leq r} \log \frac{r}{|w_k|} \\ &= \log|f(0)| + N\left(r, \frac{1}{f}\right) - N(r, f). \end{aligned}$$

For $k = 1, 2, 3, \ldots,$

$$(13.4.9) \qquad \begin{aligned} c_k(r, f) &= \frac{1}{2}\alpha_k r^k + \frac{1}{2k} \sum_{|z_n| \leq r} \left[\left(\frac{r}{z_n}\right)^k - \left(\frac{\bar{z}_n}{r}\right)^k \right] \\ &\quad - \frac{1}{2k} \sum_{|w_n| \leq r} \left[\left(\frac{r}{w_n}\right)^k - \left(\frac{\bar{w}_n}{r}\right)^k \right]. \end{aligned}$$

For $k = 1, 2, 3, \ldots,$

$$(13.4.10) \qquad c_{-k}(r, f) = \overline{c_k(r, f)}.$$

There are appropriate modifications if $f(0) = 0$ or $f(0) = \infty$.

Remark. Observe that in the notation of §1 and §2, formula (13.4.9) becomes

$$(13.4.11) \qquad \begin{aligned} c_k(r, f) &= \frac{1}{2}\alpha_k r^k + \frac{r^k}{2}\{S(r; k : Z(f)) - S(r; k : W(f))\} \\ &\quad - \frac{1}{2}\{S'(r; k : Z(f)) - S'(r; k : W(f))\}. \end{aligned}$$

Proof. We may suppose that f is holomorphic, since the result for meromorphic functions then will follow by writing f as the quotient of two holomorphic functions. We may suppose further that f has no zeros on $\{z : |z| = r\}$, since the general case follows from the continuity of both sides of (13.4.8) and (13.4.9) as functions of r. Formula (13.4.8) is, of course, Jensen's Theorem, and (13.4.10) is trivial since $\log|f|$ is real. To prove (13.4.9), write

$$I_k(r) = \frac{1}{\pi} \int_{-\pi}^{\pi} [\log f(re^{i\theta})] \cos(k\theta) \, d\theta$$

for some determination of the logarithm, and $k = 1, 2, 3, \ldots$. Then, by integrating by parts, we have

$$I_k(r) = -\frac{1}{\pi k} \int_{-\pi}^{\pi} \frac{f'(re^{i\theta})}{f(re^{i\theta})} \sin(k\theta) ire^{i\theta} \, d\theta.$$

This may be rewritten as

$$I_k(r) = \frac{1}{2\pi k i} \int_{|z|=r} \frac{f'(z)}{f(z)} \left\{ \frac{r^k}{z^k} - \frac{z^k}{r^k} \right\} \, dz.$$

This last integral may be evaluated as a sum of residues, and on taking real parts we get the kth cosine coefficient of $\log |f|$. Similarly, considering the integral

$$J_k(r) = \frac{1}{k} \int [\log(f(re^{i\theta}))] \sin(k\theta) \, d\theta,$$

where the integration is now between $\pi/2k$ and $2\pi + \pi/2k$, we get the kth sine coefficient. On combining these, we get (13.4.9).

We now define the classes of functions that we shall study.

Definition 13.4.2. Let λ be a growth function. We say that $f(z)$ is of finite λ-type and write $f \in \Lambda$ to mean that f is meromorphic and that $T(r, f) \leq A\lambda(Br)$ for some constants A, B and all positive r.

Definition 13.4.3. We denote by Λ_E the class of *all entire functions of finite λ-type*.

Proposition 13.4.4. *Let f be entire. Then f is of finite λ-type if and only if*
$\log M(r, f) \leq A\lambda(Br)$ *for some constants A, B and all positive r.*

Proof. This follows immediately from (13.4.6).

Note in particular that, if $\lambda(r) = r^\rho$, then $f \in \Lambda$ if and only if f is of growth at most order ρ, exponential-type.

We also note that by inequalities (13.4.2) and (13.4.4), Λ is a field and Λ_E is an integral domain under the usual operations.

The main theorem of this chapter is the following.

Theorem 13.4.5. *Let f be a meromorphic function. If f is of finite λ-type, then $Z(f)$ and $W(f)$ have finite λ-density and there exist constants A, B such that*

$$(13.4.12) \qquad |c_k(r, f)| \leq \frac{A\lambda(Br)}{|k| + 1} \quad (k = 0, \pm 1, \pm 2, \ldots).$$

In order that f should be of finite λ-type, it is sufficient that $Z(f)$ [or $W(f)$] have finite λ-density and that the weaker inequality

(13.4.13) $$|c_k(r, f)| \leq A\lambda(Br)$$

hold for some (possibly different) *constants A, B. Thus, in order that f should be of finite λ-type, it is necessary and sufficient that $Z(f)$ have finite λ-density and that (13.4.12) should hold. It is also necessary and sufficient that $Z(f)$ have finite λ-density and that (13.4.13) should hold.*

Proof. The order of the steps in the proof will be as follows. We first show that if f satisfies an inequality of the form (13.4.13), and if either $Z(f)$ or $W(f)$ has finite λ-density, then f must satisfy an inequality of the form (13.4.12). We then show that if f is of finite λ-type, then $Z(f)$ and $W(f)$ are of finite λ-density, and f satisfies an inequality of the form (13.4.13). Finally, we prove that if $Z(f)$ [or $W(f)$] has finite λ-density and if f satisfies an inequality of the form (13.4.12), then f must be of finite λ-type.

We shall suppose that $f(0) = 1$. The case $f(0) = 0$ or $f(0) = \infty$ causes no difficulty since we may multiply f by an appropriate power of z and the resulting function still will be of finite λ-type. This is because if $\liminf(\lambda(r)/\log r) = 0$ as $r \to \infty$, then by the exercise in Chapter 8 the class Λ contains only the constants.

Let us suppose that either $Z(f)$ or $W(f)$ is of finite λ-density and that $|c_k(r, f)| = O(\lambda(O(r)))$ uniformly for $k = 0, \pm 1, \pm 2, \ldots$. On considering the case $k = 0$, we see that both $Z(f)$ and $W(f)$ have finite λ-density. It is enough to prove that f satisfies an inequality of the form (13.4.12) for $k = 1, 2, 3, \ldots$, since c_{-k} is the complex conjugate of c_k. We prove this exactly as Proposition 13.2.6 was proved. From (13.4.11), we have

(13.4.14)
$$|c_k(r, f)| \leq \frac{r^k}{2}|\alpha_k + S(r; k : Z(f)) - S(r; k : W(f))|$$
$$+ \frac{1}{2}|S'(r; k : Z(f))| + \frac{1}{2}|S'(f; k : W(f))|,$$

and

(13.4.15)
$$\frac{r^k}{2}|\alpha_k + S(r; k : Z(f)) - S(r; k : W(f))| \leq |c_k(r, f)|$$
$$+ \frac{1}{2}|S'(r; k : Z(f))| + \frac{1}{2}|S'(f; k : W(f))|.$$

Then, by Proposition 13.2.2, for $Z = Z(f)$ or $Z = W(f)$, we have

$$|S'(r; k : Z)| = k^{-1}O(\lambda(O(r))) \quad \text{uniformly for } k > 0.$$

By (13.4.14), it is therefore enough to prove that

$$|\alpha_k + S(r; k : Z(f)) - S(r; k : W(f))| = k^{-1}r^{-k}O(\lambda(O(r)))$$
$$\text{uniformly for } k = 1, 2, 3, \ldots.$$

But, we already have from (13.4.15) that

$$|\alpha_k + S(r; k : Z(f)) - S(r; k : W(f))| = r^{-k}O(\lambda(O(r)))$$

uniformly for such k. Replacing r by $r' = k^{1/k}r$ and observing that $r' \leq 2r$, we have that

$$|\alpha_k + S(r'; k : Z(f)) - S(r'; k : W(f))| = k^{-1}r^{-k}O(\lambda(O(r))).$$

Thus, the assertion will be proved if we can show that, for $Z = Z(f)$ and $Z = W(f)$, we have

$$|S(r, r'; k : Z)| = k^{-1}r^{-k}O(\lambda(O(r))).$$

This was proved in Proposition 13.1.11 [see (13.1.3)].

Now suppose that f has finite λ-type. Then

$$N(r, W(f)) = N(r, f) \leq T(r, f),$$

so that $W(f)$ has finite λ-density. By (13.4.1), the function $1/f$ also has finite λ-type. Hence, $Z(f) = W(1/f)$ also has finite λ-density. To see that an inequality of the form (13.4.13) holds, note that

$$\begin{aligned}
|c_k(r, f)| &= \left|\frac{1}{2\pi}\int_{-\pi}^{\pi}\{\log|f(re^{i\theta})|\}e^{-ik\theta}\,d\theta\right| \\
&\leq \frac{1}{2\pi}\int_{-\pi}^{\pi}\left|\log|f(re^{i\theta})|\right|\,d\theta \leq 2T(r, f) + \log|f(0)|
\end{aligned}$$

by (13.4.4).

Finally, suppose that $W(f)$ has finite λ-density and that (13.4.12) holds. If $Z(f)$ has finite λ-density, we apply the argument below to the function $\frac{1}{f}$. Then $N(r, f) = O(\lambda(O(r)))$. It remains to prove that $m(r, f) = O(\lambda(O(r)))$. However,

$$m(r, f) \leq \frac{1}{2\pi}\int_{-\pi}^{\pi}\left|\log|f(re^{i\theta})|\right|\,d\theta,$$

which, by the Schwarz inequality, does not exceed

$$\left(\frac{1}{2\pi}\int_{-\pi}^{\pi}\left|\log|f(re^{i\theta})|\right|^2\,d\theta\right)^{\frac{1}{2}}.$$

By Parseval's Theorem, we have, for suitable constants A, B,

$$\begin{aligned}
\frac{1}{2\pi}\int_{-\pi}^{\pi}\left|\log|f(re^{i\theta})|\right|^2\,d\theta &= \sum_{k=-\infty}^{\infty}|c_k(r, f)|^2 \\
&\leq A^2(\lambda(Br))^2\sum_{k=-\infty}^{\infty}\left(\frac{1}{|k|+1}\right)^2.
\end{aligned}$$

Hence, $m(r, f) = O(\lambda(O(r)))$, which completes the proof of the theorem.

Specializing Theorem 13.4.5 to entire functions, we have the next result.

Theorem 13.4.6. *Let f be an entire function. If f is of finite λ-type, then there exist constants A, B such that*

$$(13.4.16) \qquad |c_k(r, f)| \leq \frac{A(\lambda(Br))}{|k| + 1} \quad (k = 0, \pm 1, \pm 2, \dots).$$

It is sufficient, in order that f be of finite λ-type, that there exist (possibly different) *A, B such that*

$$(13.4.17) \qquad |c_k(r, f)| \leq A\lambda(Br) \quad (k = 0, \pm 1, \pm 2, \dots).$$

Thus, in order that f should be of finite λ-type, it is necessary and sufficient that (13.4.16) should hold, and it is also necessary and sufficient that (13.4.17) should hold.

Proof. This result is an immediate corollary of Theorem 4.6 since $W(f)$ is empty in case f is entire.

Definition 13.4.7. For a meromorphic function f, we define

$$E_q(r, f) = \left\{ \frac{1}{2\pi} \int_{-\pi}^{\pi} \left| \log |f(re^{i\theta})| \right|^q d\theta \right\}^{\frac{1}{q}}.$$

Notice that if f is entire with $f(0) = 1$, and if $\alpha = \{\alpha_k\}$ is such that $c_k(r, f) = c_k(r; Z(f) : \alpha)$, $k = 0, \pm 1, \pm 2, \dots$, then $E_2(r, f) = E_2(r; Z(f) : \alpha)$, where this last quantity is the one defined in Definition 13.2.7.

Theorem 13.4.8. *Let f be an entire function. If f is of finite λ-type and $1 \leq q < \infty$, then*

$$(13.4.18) \qquad E_q(r, f) \leq A\lambda(Br)$$

for suitable constants A, B and all $r > 0$.

Conversely, if (13.4.18) holds for some $q \geq 1$, then f is of finite λ-type.

Proof. If f is of finite λ-type, then by the Hausdorff-Young Theorem ([51], p. 190), the L^q norm of $\log |f(re^{i\theta})|$, as a function of θ, is bounded by the ℓ^p norm of the sequence $\{c_k\}$, where $(\frac{1}{p}) + (\frac{1}{q}) = 1$. By Theorem 13.4.6, this ℓ^p norm is dominated by an expression of the form $A\lambda(Br)$. Conversely, using Hölder's inequality,

$$|c_k(r, f)| \leq \frac{1}{2\pi} \int_{\pi}^{\pi} \left| \log |f(re^{i\theta})| \right| d\theta,$$

$$\leq \frac{C}{2\pi} \left\{ \int_{\pi}^{\pi} \left| \log |f(re^{i\theta})| \right|^q d\theta \right\}^{\frac{1}{q}} \leq A\lambda(Br)$$

for suitable constants A, B, and it follows from Theorem 13.4.6 that f has finite λ-type.

Theorem 13.4.9. *Let f be a meromorphic function of finite λ-type, with $f(0) \neq 0, \infty$. Then for each positive number ϵ there exist positive constants α, β such that, for all $r > 0$,*

$$(13.4.19) \qquad \frac{1}{2\pi} \int_{-\pi}^{\pi} \exp\left(\frac{\alpha}{\lambda(\beta r)} \left| \log |f(re^{i\theta})| \right| \right) \, d\theta \leq 1 + \epsilon.$$

Remark. We have as a consequence that, for all $r > 0$,

$$\frac{1}{2\pi} \int_{\pi}^{\pi} \frac{1}{|f(re^{i\theta})|^{\alpha/\lambda(Br)}} \, d\theta \leq 1 + \epsilon,$$

which is somewhat surprising, even in case f is entire, since it is by no means evident that the integral is even finite.

Proof. There is a number $\beta > 0$ such that

$$\frac{|c_k(r, f)|}{\lambda(\beta r)} \leq \frac{M}{|k| + 1} \quad (k = 0, \pm 1, \pm 2, \dots).$$

Let

$$F(\theta) = F(\theta, r) = \frac{1}{\lambda(\beta r)} \log |f(re^{i\theta})|.$$

Then

$$F(\theta) = \sum \gamma_k e^{ik\theta}, \quad \text{where} \quad \gamma_k = \frac{c_k(r, f)}{\lambda(\beta r)}.$$

We may also suppose that the constant M satisfies

$$\frac{1}{2\pi} \int_{-\pi}^{\pi} |F(\theta)| \, d\theta \leq M$$

by Theorem 4.9. By a slight modification of [49] (p. 234, Example 4), we know that for any such F there exists a constant $\alpha > 0$, where α depends only on M and ϵ, such that

$$\frac{1}{2\pi} \int_{-\pi}^{\pi} \exp(\alpha |F(\theta)|) \, d\theta \leq 1 + \epsilon,$$

from which (13.4.19) follows.

13.5. Applications to Entire Functions

We present in Theorem 13.5.2 a simple necessary sufficient condition on a sequence Z of complex numbers that it be the precise sequence of zeros of some entire function of finite λ-type. The condition is that Z should be λ-admissible in the sense of Definition 13.1.15. This generalizes a well-known theorem of Lindelöf (see the remarks following the proof of the

Theorem 13.5.1, for constructing an entire function with certain properties from an appropriate sequence of Fourier coefficients associated with
a sequence of complex numbers). We also prove in Theorem 13.5.4 that
λ has the property that each meromorphic function of finite λ-type is the
quotient of two entire functions of finite λ-type if and only if λ is regular in
the sense of Definition 13.3.2. Accordingly, Propositions 13.3.5 and 13.3.6
give a large class of growth functions λ for which this is the case, including
the classical case $\lambda(r) = r^\rho$. Even this case seems to be unknown.

We turn now to our first task, the construction of an entire function
f from a sequence Z and a sequence $\{c_k(r; z : \alpha)\}$ of Fourier coefficients
associated with Z. We recall that we have assumed that $Z = \{z_n\}$ is a
sequence of nonzero complex numbers such that $z_n \to \infty$ as $n \to \infty$.

Theorem 13.5.1.   *Suppose that* $\{c_k(r)\} = \{c_k(r; Z : \alpha)\}$, $k = 0, \pm 1,$
$\pm 2, \ldots,$ *is a sequence of Fourier coefficients associated with* Z *such that
for each* $r > 0$, $\sum |c_k(r)|^2 < \infty$. *Then there exists a unique entire function
f with* $Z(f) = Z$, $f(0) = 1$, *and* $c_k(r, f) = c_k(r)$ *for* $k = 0, \pm 1, \pm 2, \ldots$.

Proof. We define

$$\Phi(\rho e^{i\varphi}) = \sum_{k=-\infty}^{\infty} c_k(\rho) e^{ik\varphi}.$$

Since $\sum |c_k(\rho)|^2 < \infty$, this defines $\Phi(\rho e^{i\varphi})$ as an element of $L^2[-\pi, \pi]$
for each $\rho > 0$ by the Riesz-Fischer Theorem. For $\rho > 0$, we define the
following functions:

$$(13.5.1) \qquad B_\rho(z; z_n) = \frac{\bar{z}_n}{|z_n|} \frac{\rho(z_n - z)}{\rho^2 - \bar{z}_n z},$$

$$(13.5.2) \qquad P_\rho(z) = \prod_{|z_n| \leq \rho} B_\rho(z; z_n),$$

$$(13.5.3) \qquad K(w; z) = \frac{w + z}{w - z},$$

$$(13.5.4) \qquad Q_\rho(z) = \exp\left\{ \frac{1}{2\pi i} \int_{|w|=\rho} K(w, z) \Phi(w) \frac{dw}{w} \right\},$$

$$(13.5.5) \qquad f_\rho(z) P_\rho(z) Q_\rho(z).$$

We make the following assertions:

$(13.5.6)$

The function f_ρ is holomorphic in the disc $\{z : |z| < \rho\}$,

and its zeros there are those z_n in Z that lie in this disc.

(13.5.7) $$f_\rho(0) = 1.$$

(13.5.8)   If $r < \rho$, then $c_k(r, f) = c_k(r)$.

Now (13.5.6) is clear from the definition of f_ρ. Also,

$$f_\rho(0) = P_\rho(0)Q_\rho(0) = Q_\rho(0) \prod_{|z_n|\leq\rho} \frac{|z_n|}{\rho}.$$

However,

$$Q_\rho(0) = \exp\left\{ \frac{1}{2\pi i} \int_{|w|=\rho} \Phi(w)\frac{dw}{w} \right\} = \exp\{c_0(\rho)\} = \prod_{|z_n|\leq\rho} \frac{\rho}{|z_n|},$$

and it follows that $f_\rho(0) = 1$.

To prove (13.5.8), we see by 13.4.1 that it is enough to show that, near $z = 0$,

$$\log f_\rho(z) = \sum_{k=1}^{\infty} \alpha_k z^k,$$

where the α_k are such that $\alpha = \{\alpha_k\}$ and $c_k(r) = c_k(r; Z : \alpha)$. That is, near $z = 0$,

(13.5.9) $$\frac{f_\rho'(z)}{f_\rho(z)} = \sum_{k=1}^{\infty} k\alpha_k z^{k-1}.$$

We now make this computation. First we have that

$$\frac{B_\rho'(z; z_n)}{B_\rho(z; z_n)} = \frac{|z_n|^2 - \rho^2}{(z_n - z)(\rho^2 - \bar{z}_n z)} = \frac{\bar{z}_n}{\rho^2 - \bar{z}_n z} - \frac{1}{z_n - z}$$

$$= \sum_{k=1}^{\infty} \left(\frac{\bar{z}_n}{\rho^2}\right)^k z^{k-1} - \sum_{k=1}^{\infty} \left(\frac{1}{z_n}\right)^k z^{k-1}.$$

Thus,

$$\frac{P_\rho'(z)}{P_\rho(z)} = \sum_{k=1}^{\infty} U_{k,\rho} z^{k-1} \quad \text{near} \quad z = 0,$$

where

$$U_{k,\rho} = \sum_{|z_n|\leq\rho} \left(\frac{\bar{z}_n}{\rho^2}\right)^k - \sum_{|z_n|\leq\rho} \left(\frac{1}{z_n}\right)^k.$$

For $k = 0, 1, 2, \ldots$, we write $w = \rho e^{i\varphi}$ and $c_k(\rho)e^{ik\varphi} = \Omega_k w^k$. Then by the definition of $c_k(\rho)$ we see that

$$\Omega_0 = N(\rho, Z)$$

and

$$\Omega_k = \frac{1}{2}\alpha_k + \frac{1}{2k}\sum_{|z_n|\leq\rho}\left\{\left(\frac{1}{z_n}\right)^k - \left(\frac{\bar{z}_n}{\rho^2}\right)^k\right\} \quad (k=1,2,3,\ldots).$$

Then

$$\Phi(w) = N(\rho,Z) + \sum_{k=1}^{\infty}\{\Omega_k w^k + \bar{\Omega}_k \bar{w}^k\}$$

$$N(\rho,Z) + \sum_{k=1}^{\infty}\left\{\Omega_k w^k + \bar{\Omega}_k \rho^{2k}\left(\frac{1}{w}\right)^k\right\},$$

so that

$$\frac{1}{2\pi i}\int_{|w|=\rho}\Phi(w)K_z(w,z)\frac{dw}{w} = \frac{2}{2\pi i}\int_{|w|=\rho}\frac{\Phi(w)}{(w-z)^2}\,dw,$$

where

$$K_z(w,z) = \frac{\partial}{\partial z}K(w,z) = \frac{2w}{(w-z)^2}.$$

But

$$\frac{1}{2\pi i}\int_{|w|=\rho}\frac{w^k}{(w-z)^2}dw = kz^{k-1} \quad \text{for} \quad k=0,1,2,\ldots,$$

and

$$\frac{1}{2\pi i}\int_{|w|=\rho}\left(\frac{1}{w}\right)^k\frac{1}{(w-z)^2}\,dw = 0 \quad \text{for} \quad k=0,1,2,\ldots.$$

Hence,

$$\frac{Q_\rho'(z)}{Q_\rho(z)} = \sum_{k=1}^{\infty}V_{k,\rho}z^{k-1},$$

where

$$V_{k,\rho} = 2\Omega_k = \alpha_k + \frac{1}{k}\sum_{|z_n|\leq\rho}\left\{\left(\frac{1}{z_n}\right)^k - \left(\frac{\bar{z}_n}{\rho^2}\right)^k\right\}.$$

Hence, near $z=0$ we have

$$\frac{f_\rho'(z)}{f_\rho(z)} = \frac{P_\rho'(z)}{P_\rho(z)} + \frac{Q_\rho'(z)}{Q_\rho(z)} = \sum_{k=1}^{\infty}k\alpha_k z^{k-1},$$

and (13.5.9) is proved

It next follows from (13.5.6)–(13.5.8) that

(13.5.10)   if $\rho' > \rho$, then $f_{\rho'}$ is an analytic continuation of f_ρ.

For if we define for $|z| < \rho$

$$F(z) = \frac{f'_\rho(z)}{f_\rho(z)},$$

then

$$c_k(r, F) = c_k(r, f_{\rho'}) - c_k(r, f_\rho) = c_k(r) - c_k(r) = 0$$

for $0 \leq r < \rho$, and therefore $|F(z)| = 1$. On the other hand, $F(0) = 1$, and it follows that F is the constant function 1.

We now define the function f of Theorem 13.5.1 by setting $f(z) = f_\rho(z)$ if $\rho > |z|$. It is clear that f is entire and, by (13.5.6), that $Z(f) = Z$. Also, $f(0) = 1$, and $c_k(r, f) = c_k(r, f_\rho)$ for $\rho > r$, so that $c_k(r, f) = c_k(r)$. An argument analogous to the one used in proving (13.5.10) proves that f is unique, and the proof of the theorem is complete.

We now characterize the zero sets of entire functions of finite λ-type.

Theorem 13.5.2. *A necessary and sufficient condition that the sequence Z be the precise sequence of zeros of an entire function f of finite λ-type is that Z be λ-admissible in the sense of Definition 13.1.15, that is, that Z have finite λ-density and be λ-balanced.*

Proof. If $Z = Z(f)$ for some $f \in \Lambda_E$, then by Theorem 13.4.6 the sequence $\{c_k(r, f)\}$ is a λ-admissible sequence of Fourier coefficients associated with Z and thus Z is λ-admissible by Proposition 13.2.5. Conversely, suppose that Z is λ-admissible. Then by Proposition 13.2.5 there exists a λ-admissible sequence $\{c_k(r)\}$ associated with Z. Then by Theorem 13.5.1 there exists an entire function f with $Z = Z(f)$ and $\{c_k(r, f)\} = \{c_k(r)\}$. Then by Theorem 14.4.7 and the fact that $\{c_k(r)\}$ is λ-admissible, it follows that $f \in \Lambda_E$, and the proof is complete.

Remark. This theorem generalizes a well-known result of Lindelöf [20], which may be stated as follows.

Theorem 13.5.3. *Let Z be a sequence of compex numbers, and let $\rho > 0$ be given. If ρ is not an integer, then in order that there exist an entire function of growth at most order ρ, finite-type, it is necessary and sufficient that there exist a constant A such that $n(r, Z) \leq Ar^\rho$. If ρ is an integer, it is necessary and sufficient that both this and the following condition be satisfied for some constant B:*

$$\left| \sum_{|z_n| \leq r} \left(\frac{1}{z_n}\right)^\rho \right| \leq B.$$

This result follows immediately from Theorem 13.5.2 and the characterization of r^ρ-admissible sequences given in Proposition 13.3.3. Our result shows that, in general, the angular distribution of the sequence of zeros of a function, and not only its density, is involved in an essential way in determining the rate of growth of the function.

We turn now to the second problem of this section, that of determining when Λ is the field of quotients of the ring Λ_E. We first prove the following result.

Theorem 13.5.4. *In order that a sequence Z of complex numbers be the precise sequence of zeros of a meromorphic function of finite λ-type, it is necessary and sufficient that Z have finite λ-density.*

Proof. The necessity follows immediately from the fact that if f is a meromorphic function, then $N(r,f) \leq T(r,f)$. For the sufficiency, we remark first that the method used in proving Theorem 13.5.1 can be used to construct suitable meromorphic functions. Indeed, suppose that we are given two disjoint sequences Z, W of nonzero complex numbers with no finite limit point and constants γ_k, $k = 1, 2, 3, \ldots$, such that the coefficients defined by

$$c_0(r) = N(r,Z) - N(r,W),$$

$$c_k(r) = \frac{r^k}{2}\{\gamma_k + S(r;k:Z) - S(r;k:W)\}$$

$$- \frac{1}{2}\{S'(r;k:Z) - S'(r;k:W) \quad (k = 1,2,3,\ldots)$$

$$c_{-k}(r) = \overline{(c_k(r))} \quad (k = 1,2,3,\ldots)$$

satisfy $\sum |c_k(r)|^2 < \infty$ for every $r > 0$. Then, by defining

$$D_\rho(z;w_n) = \prod_{|w_n| \leq \rho} B_\rho(z;w_n)$$

and

$$f_\rho(z) = \frac{P_\rho(z)Q_\rho(z)}{D_\rho(z)},$$

one can show, as in Theorem 13.5.1, that the meromorphic function defined by $f(z) = f_\rho(z)$ for $\rho > |z|$ has zero sequence Z, pole sequence W, and Fourier coefficients $\{c_k(r)\}$. It is therefore enough to prove that, given a sequence Z of finite λ-density, there exist a disjoint sequence W of finite λ-density and constants γ_k, $k = 1, 2, 3, \ldots$, such that the $c_k(r)$ satisfy $|c_k(r)| \leq A\lambda(Br)$ for some constants A, B and all $r > 0$. For then, by the first part of the proof of Theorem 13.4.5, the $c_k(r)$ must satisfy the stronger inequality

$$|c_k(r)| \leq \frac{A'\lambda(B'r)}{|k|+1} \quad (r > 0)$$

for some constants A', B', so that the function f synthesized from the $c_k(r)$ must be of finite λ-type by Theorem 13.4.5.

Supposing now that $Z = \{z_n\}$ has finite λ-density, we define $W = \{w_n\}$ by $w_n = z_n + \epsilon_n$, $n = 1, 2, 3, \ldots$, where the ϵ_n are small complex numbers so chosen that $|w_n| = |z_n|$, $n = 1, 2, 3, \ldots$, all of the numbers w_n and z_k are different, and such that

$$\sum \frac{|\epsilon_n|}{|z_n|} < \lambda(0).$$

Then, $N(r, W) = N(r, Z)$ so that W has finite λ-density. Hence,

$$|S'(r; k : Z)| = k^{-1} O(\lambda(O(r)))$$

and

$$|S'(r; k : W)| = k^{-1} O(\lambda(O(r))) \quad k = 1, 2, 3, \ldots.$$

We define

$$\gamma_k = -\frac{1}{k} \sum \left\{ \left(\frac{1}{z_n}\right)^k - \left(\frac{1}{w_n}\right)^k \right\}.$$

It remains to prove that

$$\frac{r^k}{2} |\gamma_k + S(r; k : Z) - S(r; k : W)| = O(\lambda(O(r)))$$

uniformly for $k = 1, 2, 3, \ldots$. Now

$$\frac{r^k}{2} |\gamma_k + S(r; k : Z) - S(r; k : W)|$$

$$= \frac{r^k}{2} \left| \frac{1}{k} \sum_{|z_n| > r} \left\{ \left(\frac{1}{z_n}\right)^k - \left(\frac{1}{w_n}\right)^k \right\} \right|$$

$$= \frac{r^k}{2} \left| \frac{1}{k} \sum_{|z_n| > r} \frac{(w_n)^k - (z_n)^k}{(w_n z_n)^k} \right| \leq \frac{r^k}{2} \frac{1}{k} \sum_{|z_n| > r} \frac{|(w_n)^k - (z_n)^k|}{|z_n|^{2k}}.$$

However, $|(w_n)^k - (z_n)^k| \leq k |\epsilon_n| |z_n|^{k-1}$, so that we have

$$\frac{r^k}{2} |\gamma^k + S(r; k : Z) - S(r; k : W)|$$

$$\leq \frac{r^k}{2} \sum_{|z_n| > r} \frac{|\epsilon_n|}{|z_n|^{k+1}} \leq \frac{1}{2} \sum_{|z_n| > r} \frac{|\epsilon_n|}{|z_n|} \leq \frac{1}{2} \lambda(0) \leq \lambda(r).$$

Theorem 13.5.5.   *The field Λ of all meromorphic functions of finite λ-type is the field of quotients of the rings Λ_E of all entire functions of finite λ-type if and only if λ is regular in the sense of Definition 13.3.2, that is, if and only if every sequence of finite λ-density is λ-balanceable.*

Proof. First, suppose that λ is regular and that $f \in \Lambda$. Then $Z(f)$ has finite λ-density by Theorem 13.5.3. There then exists a sequence $Z' \supseteq Z(f)$ such that Z' is λ-admissible. [We may suppose, by the remarks preceding the proof of Theorem 13.4.5, that $f(0) \neq 0, \infty$]. Then, by Theorem 13.5.2, there exists a function $g \in \Lambda_E$ such that $Z(g) = Z'$. Since we have then that $Z(g) \subseteq Z(f)$, the function $h = g/f$ is entire. However,

$$T(r, h) \leq T(r, g) + T\left(r, \frac{1}{f}\right) = T(r, g) + T(r, f) - \log|f(0)|$$

by (13.4.1) and (13.4.2), so that $h \in \Lambda_E$, and $f = g/h$ is the desired representation.

Conversely, suppose that $\Lambda = \Lambda_E/\Lambda_E$. Let Z have finite λ-density. Then, by Theorem 13.5.3, there exists a function $f \in \Lambda$ with $Z(f) = Z$. We write $f = g/h$ with $g, h \in \Lambda_E$. Then $Z(g)$ is λ-admissible, and $Z(g) \supseteq Z(f) = Z$, and we have proved that λ is regular.

14

The Miles-Rubel-Taylor Theorem on Quotient Representations of Meromorphic Functions

Let f be a meromorphic function. In this chapter we describe the work of Joseph Miles, which completes the work in the last chapter concerning representations of f as the quotient of entire functions with small Nevanlinna characteristic. Miles showed that every set Z of finite λ-density is λ-balanceable. As a consequence of this and the work of Rubel and Taylor in the last chapter, there exist absolute constants A and B such that if f is any meromorphic function in the plane, then f can be expressed as f_1/f_2, where f_1 and f_2 are entire functions such that $T(r, f_i) \leq AT(Br, f)$ for $i = 1, 2$ and $r > 0$. It is implicit in the method of proof that for any $B > 1$ there is a corresponding A for which the desired representation holds for all f. Miles' proof is ingenious, intricate, and deep. Miles also showed that, in general, B cannot be chosen to be 1 by giving an example of a meromorphic f such that if $f = f_1/f_2$, where f_1 and f_2 are entire, then $T(r, f_2) \neq O(T(r, f))$. We do not give this example here.

In the previous chapter, namely in Propositions 13.3.5 and 13.3.6 using Theorem 13.5.2, we have obtained the above theorem for special classes of entire functions. Results in this direction for functions of several complex variables appear in [16], [17], and [10]. Quotient representations of functions meromorphic in the unit disk are discussed in [2]. The presentation below follows Mile [24].

We state the theorem.

Theorem. *There exists absolute constants A and B such that if f is any meromorphic function in the plane, then there exist entire functions f_1 and*

f_2 such that $f = f_1/f_2$ and such that $T(r, f_i) \leq AT(Br, f)$ for $i = 1, 2$ and $r > 0$.

Suppose $Z = \{z_n\}$ is a sequence of nonzero complex numbers with $|z_n| \to \infty$. We include the possibility that $z_n = z_m$ for some $n \neq m$. As in the previous chapter, let

$$(14.1) \qquad n(r, Z) = \sum_{|z_n| \leq r} 1$$

and

$$(14.2) \qquad N(r, Z) = \int_0^r \frac{n(t, Z)}{t} \, dt.$$

It was shown in the previous chapter that the following lemma is sufficient to establish the theorem.

Lemma. *Suppose $Z = \{z_n\}$ is a sequence of nonzero complex numbers with $|z_n| \to \infty$. If $\lambda(r) = \max(1, N(r, Z))$, then there exist absolute constants A' and B' and a sequence $\tilde{Z} = \{\tilde{z}_n\}$ containing Z (with due regard to multiplicities) such that*

(i) $\qquad\qquad\qquad N(r, \tilde{Z}) \leq 5\lambda(4r) \quad r > 0$

and,

(ii) *for $j = 1, 2, 3, \ldots$ and $s > r > 0$,*

$$\left| \frac{1}{j} \sum_{r < |\tilde{z}_n| \leq s} \left(\frac{1}{\tilde{z}_n}\right)^j \right| \leq \frac{A'\lambda(B'r)}{r^j} + \frac{A'\lambda(B's)}{s^j}.$$

The argument of the last chapter which shows that this lemma is sufficient to prove the theorem may be summarized as follows. Without loss of generality we may assume f has infinitely many poles and that $f(0) \neq \infty$. Let Z be the sequence of poles of f. Recall from the last chapter that condition (i) of the lemma says that $\tilde{Z}$ has finite density with respect to the growth function $\lambda(r)$ and condition (ii) says that $\tilde{Z}$ is balanced with respect to the growth function $\lambda(r)$. Let $\lambda_1(r)$ denote an arbitrary increasing unbounded function defined on $(0, \infty)$.

In Theorem 13.5.2 we characterized the zero sets of entire functions ϕ such that $T(r, \phi) \leq \alpha\lambda_1(\beta_r)$ for some constants α and β and all $r > 0$ as those sets Z^* which both have finite density and are balanced with respect to $\lambda_1(r)$. This characterization combined with the above lemma guarantees the existence of constants A_1 and B and of an entire function f_2 having zeros precisely on the set $\tilde{Z}$ (counting multiplicities) such that $T(r, f_2) \leq A_1\lambda_1(Br)$ for all $r > 0$. Hence,

$$(14.3) \qquad T(r, f_2) \leq A_1 N(Br, Z) \leq A_1 T(Br, f)$$

for $r > r_0(f)$. Letting $f_1 = f_2 f$, we see that f_1 is entire and that $T(r, f_1) \leq (A_1 + 1)T(Br, f)$ for $r > r_0(f)$. For an appropriate complex constant c, $0 < |c| < 1$, we have for $i = 1, 2$ that

$$(14.4) \qquad\qquad T(r, cf_i) = 0 \quad r < r_0(f)$$

and

$$(14.5) \qquad\qquad T(r, cf_i) \leq T(r, f_i) \quad r > 0.$$

Letting $A = A_1 + 1$, we see that $f = cf_1/cf_2$ is the desired representation. It is implicit in the methods of the last chapter and in the proof of the above lemma that A and B are absolute constants and that to each $B > 1$ there corresponds an A for which the representation holds for all f.

We now prove the lemma. For each integer N we let

$$Z_N = Z \cap \{z : 2^N < |z| \leq 2^{N+1}\}.$$

If $Z_N \neq \varnothing$, we relabel the elements of Z_N as simply $z_1, z_2, \ldots, z_k$, with each number being listed in this sequence as often as it appears in Z. For $1 \leq n \leq k$ we define $p_n \in (0, 1]$ and $\theta_n \in (0, 2\pi]$, so that $z_n = 2^{N+p_n} e^{i\theta_n}$. We do not indicate in the notation the obvious dependence of k, z_n, p_n, and θ_n on N. We let

$$(14.6) \qquad\qquad g_N(\theta) = -2 \sum_{n=1}^{k} \left\{ \sum_{j=1}^{\infty} \frac{e^{ij(\theta - \theta_n)}}{2^{j(1+p_n)}} \right\}$$

and

$$(14.7) \qquad\qquad h_N(\theta) = \Re g_N(\theta).$$

Clearly,

$$(14.8) \qquad\qquad |h_N(\theta)| \leq 2 \sum_{n=1}^{k} \left\{ \sum_{j=1}^{\infty} 2^{-j(1+p_n)} \right\}$$
$$< (n(2^{N+1}, Z) - n(2^N, Z)).$$

Letting

$$(14.9) \qquad f_N(\theta) = h_N(\theta) + 2(n(2^{N+1}, Z) - n(2^N, Z)),$$

we have

$$(14.10) \qquad 0 < f_N(\theta) < 4(n(2^{N+1}, Z) - n(2^N, Z)).$$

We now define sets Z'_N for all integers N. If $Z_N \neq \phi$, we have from (14.10) that

$$(14.11) \qquad 0 < \frac{1}{2\pi} \int_0^{2\pi} f_N(\theta) \, d\theta < 4(n(2^{N+1}, Z) - n(2^N, Z)).$$

If [ ] denotes the greatest integer function and $L_N = \left[\frac{1}{2\pi} \int_0^{2\pi} f_N(\theta) \, d\theta\right]$, we define for $0 \leq n \leq L_N$ a monotone increasing sequence θ'_n in $[0, 2\pi]$ by choosing θ'_n to satisfy

$$(14.12) \qquad \frac{1}{2\pi} \int_0^{\theta'_n} f_N(\theta) \, d\theta = n.$$

In addition, we let $\theta'_{L_N+1} = 2\pi$. For $0 \leq n \leq L_N$, we let $z'_n = 2^{N-1} e^{i\theta'_n}$ and $Z'_N = \{2^{N-1} e^{i\theta'_n} : 0 \leq n \leq L_N\}$. As before, we do not indicate the dependence of θ'_n and z'_n on N. If $Z_N = \varnothing$, we let $Z'_N = \varnothing$ and for that value of N do not define L_N or numbers θ'_n and z'_n.

We let $Z' = \cup_N Z'_N$ and $\tilde{Z} = Z \cup Z'$. From (14.11) and the definition of $\tilde{Z}$ it follows that $n(r, Z') \leq 4n(4r, Z)$ and hence that $n(r, \tilde{Z}) \leq 5n(4r, Z)$ for all $r > 0$. From this fact it is immediate that $\tilde{Z}$ satisfies condition (i) of the lemma.

We now consider a positive integer j and a value of N for which $Z_N \neq \phi$. Let $Z_N = \{z_1, z_2, \ldots, z_k\}$. From this point until inequality (14.27), we regard j and N as fixed. Although many of the quantities to be defined ($S, T, P, U_0, U_e, V_0,$ and V_e) depend on both j and N, for simplicity we suppress this dependence from the notation.

A key step in showing $\tilde{Z}$ satisfies condition (ii) of the lemma is to establish

$$(14.13) \qquad \left| \sum_{|z'_n|=2^{N-1}} \left(\frac{1}{z'_n}\right)^j + \sum_{n=1}^k \left(\frac{1}{z_n}\right)^j \right| < 48j2^{-j(N-1)}.$$

We first observe that
(14.14)

$$\sum_{n=1}^k \left(\frac{1}{z_n}\right)^j + \frac{1}{2\pi} \int_0^{2\pi} e^{-ij\theta} 2^{-j(N-1)} f_N(\theta) \, d\theta$$

$$= \sum_{n=1}^k \left(\frac{1}{z_n}\right)^j - \frac{1}{2\pi} \int_0^{2\pi} e^{-ij\theta} 2^{-j(N-1)} \left(\sum_{n=1}^k \frac{e^{ij(\theta-\theta_n)}}{2^{j(1+p_n)}}\right) d\theta$$

$$= \sum_{n=1}^k \left(\frac{1}{z_n}\right)^j - \sum_{n=1}^k 2^{-j(N+p_n)} e^{-ij(\theta_n)} = 0.$$

Thus (14.13) is equivalent to
(14.15)

$$\left| \sum_{|z_n'|=2^{N-1}} \left(\frac{1}{z_n'}\right)^j - \frac{1}{2\pi} \int_0^{2\pi} e^{-ij\theta} 2^{-j(N-1)} f_N(\theta)\, d\theta \right| < 48 j 2^{-j(N-1)}.$$

That the quantity on the left side of (14.15) is small can be seen intuitively from the observation that the sum is essentially an approximating Riemannn sum for the integral. We first estimate

(14.16)

$$\left| \Re \left\{ \sum_{|z_n'|=2^{N-1}} \left(\frac{1}{z_n'}\right)^j - \frac{1}{2\pi} \int_0^{2\pi} e^{-ij\theta} 2^{-j(N-1)} f_N(\theta)\, d\theta \right\} \right|$$

$$= 2^{-j(N-1)} \left| \sum_{|z_n'|=2^{N-1}} \cos j\theta_n' - \frac{1}{2\pi} \int_0^{2\pi} f_N(\theta) \cos j\theta\, d\theta \right|.$$

The function $\cos j\theta$ decreases from 1 to -1 on $\left[\frac{m\pi}{j}, \frac{(m+1)\pi}{j}\right]$ for m even, $0 \le m \le 2j - 1$, and increases from -1 to 1 on $\left[\frac{m\pi}{j}, \frac{(m+1)\pi}{j}\right]$ for m odd, $0 \le m \le 2j - 1$. Let $I_m = \left(\frac{m\pi}{j}, \frac{(m+1)\pi}{j}\right)$ for $0 \le m \le 2j - 1$. Define S to be the set of all n, $0 \le n \le L_N$, such that $[\theta_n', \theta_{n+1}']$ is not contained entirely in some I_m, $0 \le m \le 2j - 1$. Since each point $m\pi/j$ belongs to $[\theta_n', \theta_{n+1}']$ for at most two values of n, we see that S has at most $4j$ elements. Let T be the set of all n, $0 \le n \le L_N$, such that either n or $n-1$ belongs to S. Thus T has at most $8j$ elements.

Let $P = \{0, 1, \ldots, L_N\} - T$. If $n \in P$, it follows that $[\theta_{n+1}', \theta_n']$ and $[\theta_n', \theta_{n+1}']$ belong to the same interval I_m for some m, $0 \le m \le 2j - 1$. Let U_o be the set of all $n \in P$ such that $[\theta_{n+1}', \theta_n']$ and $[\theta_n', \theta_{n+1}']$ are contained in I_m for some odd m, and let U_e be the set of all $n-1$ such that $n \in P$ and the intervals $[\theta_{n-1}', \theta_n']$ and $[\theta_n', \theta_{n+1}']$ are contained in I_m for some even m. Certainly $U_o \cap U_e = \varnothing$, for if $n \in U_o \cap U_e$, then $[\theta_n', \theta_{n+1}']$ is contained in I_m for both an odd value of m and an even value of m. Letting $U = U_o \cup U_e$, we conclude that U and P have the same number of elements. Since $0 \notin P$, we see that $-1 \notin U_e$ and hence $U \subset \{0, 1, \ldots, L_N\}$. Thus, $\{0, 1, \ldots, L_N\} - U$ has at most $8j$ elements.

If $n \in P$ is such that $u \in U_o$, then $\cos j\theta$ is increasing on $[\theta_n', \theta_{n+1}']$ and hence, from (14.12),

(14.17)
$$\cos j\theta_n' \le \frac{1}{2\pi} \int_{\theta_n'}^{\theta_{n+1}'} f_N(\theta) \cos j\theta\, d\theta.$$

If $n \in P$ is such that $n-1 \in U_e$, then $\cos j\theta$ is decreasing on $[\theta_{n+1}', \theta_n']$ and thus

(14.18)
$$\cos j\theta_n' \le \frac{1}{2\pi} \int_{\theta_{n-1}'}^{\theta_n'} f_N(\theta) \cos j\theta\, d\theta.$$

The definition of U together with (14.17) and (14.18) implies

(14.19)
$$\sum_{n \in P} \cos j\theta'_n \leq \sum_{n \in U} \frac{1}{2\pi} \int_{\theta'_n}^{\theta'_{n+1}} f_N(\theta) \cos j\theta \, d\theta.$$

It follows that
(14.20)
$$\sum_{|z'_n|=2^{N-1}} \cos j\theta'_n - \frac{1}{2\pi} \int_0^{2\pi} f_N(\theta) \cos j\theta \, d\theta$$
$$= \sum_{n \in P} \cos j\theta'_n + \sum_{n \in T} \cos j\theta'_n$$
$$- \sum_{n \in U} \frac{1}{2\pi} \int_{\theta'_n}^{\theta'_{n+1}} f_N(\theta) \cos j\theta \, d\theta - \sum_{\substack{0 \leq n \leq L_N \\ n \notin U}} \frac{1}{2\pi} \int_{\theta'_n}^{\theta'_{n+1}} f_N(\theta) \cos j\theta \, d\theta$$
$$\leq \sum_{n \in T} \cos j\theta'_n - \sum_{\substack{0 \leq n \leq L_N \\ n \notin U}} \frac{1}{2\pi} \int_{\theta'_n}^{\theta'_{n+1}} f_N(\theta) \cos j\theta \, d\theta$$
$$\leq 8j + 8j = 16j.$$

We now obtain a lower bound for

(14.21)
$$\sum_{|z'_n|=2^{N-1}} \cos j\theta'_n - \frac{1}{2\pi} \int_0^{2\pi} f_N(\theta) \cos j\theta \, d\theta.$$

Let V_e be the set of all $n \in P$ such that $[\theta'_{n-1}, \theta'_n]$ and $[\theta'_n, \theta'_{n+1}]$ are contained in I_m for some even m, and let V_o be the set of all $n-1$ such that $n \in P$ and the intervals $[\theta'_{n-1}, \theta'_n]$ and $[\theta'_n, \theta'_{n+1}]$ are contained in I_m for some odd m. Using the same reasoning as before, we see that $V_e \cap V_o = \emptyset$ and hence, if $V = V_e \cup V_o$, then $\{0, 1, \ldots, L_N\} - V$ has at most $8j$ elements.

If $n \in P$ is such that $n \in V_e$, then $\cos j\theta$ is decreasing on $[\theta'_n, \theta'_{n+1}]$ and

(14.22)
$$\cos j\theta'_n \geq \frac{1}{2\pi} \int_{\theta'_n}^{\theta'_{n+1}} f_N(\theta) \cos j\theta \, d\theta.$$

If $n \in P$ is such that $n - 1 \in V_o$, then $\cos j\theta$ is increasing on $[\theta'_{n-1}, \theta'_n]$ and

(14.23)
$$\cos j\theta'_n \geq \frac{1}{2\pi} \int_{\theta'_{n-1}}^{\theta'_n} f_N(\theta) \cos j\theta \, d\theta.$$

From (14.22), (14.23), and the definition of V,

(14.24)
$$\sum_{n \in P} \cos j\theta'_n \geq \sum_{n \in V} \frac{1}{2\pi} \int_{\theta'_n}^{\theta'_{n+1}} f_N(\theta) \cos j\theta \, d\theta.$$

Thus
(14.25)

$$\sum_{|z'_n|=2^{N-1}} \cos j\theta'_n - \frac{1}{2\pi}\int_0^{2\pi} f_N(\theta)\cos j\theta \, d\theta$$

$$= \sum_{n\in P} \cos j\theta'_n + \sum_{n\in T} \cos j\theta'_n$$

$$- \sum_{n\in V} \frac{1}{2\pi}\int_{\theta'_n}^{\theta'_{n+1}} f_N(\theta)\cos j\theta \, d\theta - \sum_{\substack{0\leq n\leq L_N \\ n\notin V}} \frac{1}{2\pi}\int_{\theta'_n}^{\theta'_{n+1}} f_N(\theta)\cos j\theta \, d\theta$$

$$\geq \sum_{n\in T} \cos j\theta'_n - \sum_{\substack{0\leq n\leq L_N \\ n\notin V}} \frac{1}{2\pi}\int_{\theta'_n}^{\theta'_{n+1}} f_N(\theta)\cos j\theta \, d\theta$$

$$\geq -8j - 8j = -16j.$$

Combining (14.16), (14.20), and (14.25), we conclude that
(14.26)

$$\left|\Re\left\{\sum_{|z'_n|=2^{N-1}} \left(\frac{1}{z'_n}\right)^j - \frac{1}{2\pi}\int_0^{2\pi} e^{-ij\theta} 2^{-j(N-1)} f_N(\theta)\, d\theta\right\}\right| \leq 16j2^{-j(N-1)}.$$

The same discussion applies to the imaginary part of the above quantity. The only minor modification is that we must divide $[0,2\pi]$ into $2j+1$ subintervals on which $\sin j\theta$ is alternately increasing and decreasing. Since $2j+1 < 4j$, this causes no difficulty. We obtain
(14.27)

$$\left|\Im\left\{\sum_{|z'_n|=2^{N-1}} \left(\frac{1}{z'_n}\right)^j - \frac{1}{2\pi}\int_0^{2\pi} e^{-ij\theta} 2^{-j(N-1)} f_N(\theta)\, d\theta\right\}\right| \leq 32j2^{-j(N-1)}.$$

Combining (14.26) and (14.27), we obtain (14.15), which in turn establishes (14.13).

We are now in position to show that $\tilde{Z}$ satisfies condition (ii) of the lemma. Suppose that $s \leq 8r$. We then have trivially for all positive integers j

(14.28)
$$\left|\frac{1}{j}\sum_{r<|\tilde{z}'_n|\leq s} \left(\frac{1}{\tilde{z}'_n}\right)^j\right| \leq \frac{n(s,\tilde{Z})}{jr^j}$$

$$\leq \frac{n(8r,\tilde{Z})}{jr^j} \leq \frac{N(8er,\tilde{Z})}{jr^j} \leq \frac{5\lambda(32er)}{jr^j}.$$

Suppose that $8r < s$. In this case there exist integers q_1 and q_2, $q_1 \leq q_2 - 3$, such that

(14.29) $2^{q_1} \leq r < 2^{q_1+1} < \cdots < 2^{q_2} \leq s < 2^{q_2+1}.$

Thus, for $j = 1, 2, 3, \ldots,$

(14.30)

$$\left| \frac{1}{j} \sum_{r < |\bar{z}_n| \le s} \left(\frac{1}{\bar{z}_n}\right)^j \right| = \left| \frac{1}{j} \sum_{r < |z_n| \le 2^{q_1}+2} \left(\frac{1}{\bar{z}_n}\right)^j \right.$$

$$+ \sum_{\nu=2}^{q_2-q_1-1} \left\{ \frac{1}{j} \sum_{2^{q_1}+\nu < |z_n| \le 2^{q_1}+\nu+1} \left(\frac{1}{z_n}\right)^j + \frac{1}{j} \sum_{|z'_n| = 2^{q_1}+\nu-1} \left(\frac{1}{z_n}\right)^j \right\}$$

$$+ \left\{ \frac{1}{j} \sum_{2^{q_2} < |z_n| \le s} \left(\frac{1}{z_n}\right)^j + \frac{1}{j} \sum_{|z'_n| = 2^{q_2}-1} \left(\frac{1}{z'_n}\right)^j \right\}$$

$$+ \left. \left\{ \frac{1}{j} \sum_{|z'_n| = 2^{q_2}} \left(\frac{1}{z'_n}\right)^j \right\} \right|.$$

We remark that if $Z_{q_1+\nu} = \varnothing$ (and hence $Z'_{q_1+\nu} = \varnothing$) for some integer ν, then the terms corresponding to that value of ν in (14.30) are of course omitted.

For any positive integer j we have

(14.31)

$$\left| \frac{1}{j} \sum_{r < |z_n| \le 2^{q_1}+2} \left(\frac{1}{\bar{z}_n}\right)^j \right| \le \frac{n(4r, Z)}{jr^j}$$

$$\le \frac{N(4er, Z)}{jr^j} \le \frac{\lambda(4er)}{jr^j}.$$

From (14.13) we have

(14.32)

$$\sum_{\nu=2}^{q_2-q_1-1} \left| \frac{1}{j} \sum_{2^{q_1}+\nu < |z_n| \le 2^{q_1}+\nu+1} \left(\frac{1}{z_n}\right)^j + \frac{1}{j} \sum_{|z'_n| = 2^{q_1}+\nu-1} \left(\frac{1}{z_n}\right)^j \right|$$

$$\le 48 \sum_{\nu=2}^{q_2-q_1-1} 2^{-j(q_1+\nu-1)}.$$

Also from (14.13) we have

(14.33)

$$\left| \frac{1}{j} \sum_{2^{q_2} < |z_n| \le s} \left(\frac{1}{z_n}\right)^j + \frac{1}{j} \sum_{|z'_n| = 2^{q_2}-1} \left(\frac{1}{z'_n}\right)^j \right|$$

$$\left| \frac{1}{j} \sum_{2^{q_2} < |z_n| \le 2^{q_2}+1} \left(\frac{1}{z_n}\right)^j + \frac{1}{j} \sum_{|z'_n| = 2^{q_2}-1} \left(\frac{1}{z_n}\right)^j - \frac{1}{j} \sum_{s < |z_n| \le 2^{q_2}+1} \left(\frac{1}{z_n}\right)^j \right|$$

$$\le (48)2^{-j(q_2-1)} + \frac{n(2^{q_2+1}, Z)}{js^j} \le (48)2^{-j(q_2-1)} + \frac{\lambda(2es)}{js^j}.$$

Finally,
(14.34)
$$\left| \frac{1}{j} \sum_{|z'_n|=2^{q_2}} \left(\frac{1}{z'_n}\right)^j \right| =$$

$$\left| \frac{1}{j} \sum_{2^{q_2+1} < |z_n| \le 2^{q_2+2}} \left(\frac{1}{z_n}\right)^j + \frac{1}{j} \sum_{|z'_n|=2^{q_2}} \left(\frac{1}{z'_n}\right)^j - \frac{1}{j} \sum_{2^{q_2+1} < |z_n| \le 2^{q_2+2}} \left(\frac{1}{z_n}\right)^j \right|$$

$$\le (48)2^{-j(q_2)} + \frac{n(2^{q_2+2}, Z)}{js^j} \le (48)2^{-j(q_2)} + \frac{\lambda(4es)}{js^j}.$$

Combining (14.30) through (14.34), we have

$$\left| \frac{1}{j} \sum_{r < |\tilde{z}_n| \le s} \left(\frac{1}{\tilde{z}_n}\right)^j \right|$$

(14.35)
$$\le \frac{\lambda(4er)}{jr^j} + (48) \sum_{\nu=2}^{q_2-q_1+1} 2^{-j(q_1+\nu-1)} + 2\frac{\lambda(4es)}{js^j}$$

$$\le \frac{\lambda(4er)}{jr^j} + \frac{96}{2^{j(q_1+1)}} + \frac{2\lambda(4es)}{js^j} \le \frac{\lambda(4er)}{jr^j} + \frac{96\lambda(r)}{r^j} + \frac{2\lambda(4es)}{js^j}.$$

From (14.28) and (14.35), we see that

(14.36)
$$\left| \frac{1}{j} \sum_{r < |\tilde{z}_n| \le s} \left(\frac{1}{\tilde{z}_n}\right)^j \right| \le \frac{A'\lambda(B'r)}{r^j} + \frac{A'\lambda(B's)}{s^j}$$

for all positive integers j and all $0 < r < s$ if $A' = 100$ and $B' = 32e$. This completes the proof of the lemma.

15
Canonical Products

We shall suppose that a countable set $Z = \{z_n\}$ of complex numbers is given. For convenience, we shall suppose that $0 \notin Z$ and that Z has no finite limit points. More generally, we consider "sets with multiplicity." This means that some of the z_n may be counted multiple times. It is possible to make this notion rigorous, but at the price of clumsier notation. For convenience in this chapter, we shall exclude the null function $F(z) = 0$ from consideration.

Definition. If F is an entire function, we write $Z(f)$ for the set of zeros (other than the origin) of F.

Definition. If p is a nonnegative integer, we define $E(u, p)$, the *Weierstrass primary factor of order p*, by

$$E(u, p) = (1 - u) \exp\left(u + \frac{u^2}{2} + \cdots + \frac{u^p}{p} \right)$$

$$E(u, 0) = 1 - u.$$

Lemma 15.1. *Given any Z, there exist nonnegative integers λ_n such that the product*

$$f(z) = \prod_{n=1}^{\infty} E\left(\frac{z}{z_n}, \lambda_n \right)$$

converges uniformly on compact sets to an entire function f such that $Z(f) = Z$.

We omit the easy proof which follows directly from Estimate A later in this chapter. Any sequence λ_n such that $\lambda_n \to \infty$ will work. The next theorem is a corollary of Lemma 15.1.

Weierstrass Factorization Theorem. *Given an entire function F, there exist nonnegative integers λ_n, a nonnegative integer m, and an entire function g such that*

$$F(z) = z^m \exp(g(z)) \prod_{n=1}^{\infty} E\left(\frac{z}{z_n}, \lambda_n\right).$$

Definition. Given any Z, the *convergence exponent* $\rho_1(Z)$ is defined by

$$\rho_1(Z) = \inf\{\alpha : \Sigma|z_n|^{-\alpha} < \infty\}.$$

Lemma 15.2. *If $\Sigma \frac{1}{|z_n|^\alpha} < \infty$, then $n(t) = o(t^\alpha)$, where $n(t) = \displaystyle\sum_{|z_n| \leq t} 1$.*

Proof. Choose a large number a, and write

$$\sum_{a \leq |z_n| \leq t} \frac{1}{|z_n|^\alpha} \geq \frac{n(t) - n(a)}{t^\alpha}.$$

Lemma 15.3. *$\Sigma \frac{1}{|z_n|^\alpha} < \infty$ if and only if $\int_0^\infty \frac{n(t)}{t^{\alpha+1}} \, dt < \infty$.*

Proof. Write

$$\sum_{|z_n| < x} \frac{1}{|z_n|^\alpha} = \int_0^x t^{-\alpha} \, dn(t)$$

and integrate by parts to prove the lemma.

Corollary. $\rho_1(Z) = \inf\{\alpha : n(t) = O(t^\alpha)\}$.

Lemma 15.4. $\rho_1(f) \leq \rho(f)$.

Proof. We must show that if $\rho = \rho(f)$, then for each $\epsilon > 0$, $n(t) = O(t^{\rho+\epsilon})$. But by the Nevanlinna first fundamental theorem, we know that

$$N(t) = \int_0^t \frac{n(s)}{s} \, ds = O(t^{\rho+\epsilon}).$$

But $N(t) \geq n\left(\frac{t}{2}\right)\log 2$, by the usual argument, and the result follows.

Definition. The *genus of Z*, $p = p(Z)$, is defined by $p(Z) = \inf\{q : q$ is an integer, $\Sigma \frac{1}{|z_n|^{q+1}} < \infty\}$.

Definition. The *canonical product of genus p over Z* is defined by

$$P(z) = \prod_{z_n \in Z} E\left(\frac{z}{z_n}, p\right).$$

Definition. The canonical product P_Z over Z is defined by

$$P_Z(z) = \prod_{z_n \in Z} E\left(\frac{z}{z_n}, p\right),$$

where $p = p(Z)$.

Theorem 15.5. *The order of P_Z is $\rho_1(Z)$.*

The next result is a corollary of Theorem 15.5.

The Hadamard Factorization Theorem. *Given an entire function f,*

$$f(z) = P_Z(z) z^m \exp Q(z),$$

where m is a nonnegative integer and Q is a polynomial of degree $\leq \rho(f)$. Here, $Z = Z(f)$.

Definition. The *genus of the entire function f* is defined by

$$P(f) = \max(n, p),$$

where $n = \deg Q$.

Examples. Suppose $z_n = n^2$, $n = 1, 2, 3, \ldots$. If $f(z) = \prod \left(1 - \frac{z}{n^2}\right)$, then f is of genus 0. If $f(z) = e^z \prod \left(1 - \frac{z}{n^2}\right)$, then f is of genus 1.

We first shall show how the Hadamard Factorization Theorem follows from Theorem 15.5. Without loss of generality, suppose $f(0) \neq 0$. In fact, assume $f(0) = 1$. Let

$$g(z) = \frac{f(z)}{P_Z(z)}.$$

Then g is an entire function without zeros, so that there exists an entire function Q such that

$$\frac{f}{P_Z} = \exp Q.$$

We must prove that Q is a polynomial of degree $\leq \rho(f)$.

Lemma 15.6. *If F and G are meromorphic, then $\rho\left(\frac{F}{G}\right) \leq \max\{\rho(F), \rho(G)\}$.*

Proof. We have

$$\rho(f) = \limsup_{r \to \infty} \frac{\log T(r, f)}{\log r}$$

$$T\left(r, \frac{1}{f}\right) = T(r, f) + O(1)$$

$$T(r, fg) = T(r, f) + T(r, g) + O(1),$$

from which the lemma is easily proved.

It then follows that $\exp Q$ is an entire function of order at most ρ. Writing $Q = u + iv$, we have

$$\int_{-\pi}^{\pi} |u(re^{i\theta})| \, d\theta = O(r^{\rho'})$$

for each $\rho' > \rho$. Since, with $|z| = r$ and $R = 2r$, we have

$$Q(z) - Im \ Q(0) = \frac{1}{2\pi i} \int_{|w|=R} u(w) \frac{w+z}{w-z} \frac{dw}{w},$$

it follows that

$$|Q(z) - Im \ Q(0)| = O(r^{\rho'}).$$

Hence

$$|Q(z)| = O(r^{\rho'}),$$

and it follows that Q is a polynomial of degree $\leq \rho$.

Proof of Theorem 15.5. Our proof uses the Fourier series method of Chapter 13. It is shorter and less tedious than the standard proofs.

Estimate A. If $|z_n| \geq 2|z|$, then

$$\left| \log E\left(\frac{z}{z_n}, p\right) \right| \leq 2\left|\frac{z}{z_n}\right|^{p+1} \leq \left|\frac{z}{z_n}\right|^p.$$

Let $u = \frac{z}{z_n}$. Now $\log E(u, p) = -\sum_{p+1}^{\infty} \frac{u^k}{k}$. Here $|u| \leq \frac{1}{2}$. Hence

$$|\log E(u, p)| \leq \sum_{p+1}^{\infty} |u|^k \leq |u|^{p+1} \sum_{0}^{\infty} 2^{-k} = 2|u|^{p+1}.$$

It follows from Estimate A that the product

$$P_Z(z) = \Pi E\left(\frac{z}{z_n}, p\right)$$

converges uniformly on compact sets to an entire function f, so that $\log |f(z)| = \sum \log \left| E\left(\frac{z}{z_n}, p\right) \right|$. We write

$$\log |f(z)| = \sum_{m=-\infty}^{\infty} c_m \, e^{im\theta},$$

where $c_m = c_m(r)$ is the mth Fourier coefficient of $\log|f|$. We have seen in Lemma 13.4.1 how to calculate c_m:

$$\log E\left(\frac{z}{z_n},\ P\right) = -\frac{1}{k}\sum_{k=p+1}^{\infty}\left(\frac{z}{z_n}\right)^k$$

so that

$$\log f(z) = \sum_{n=0}^{\infty}\left(-\sum_{k=p+1}^{\infty}\frac{1}{k}\left(\frac{z}{z_n}\right)^k\right).$$

Hence

$$\frac{1}{2}\alpha_n = \begin{cases} 0 & \text{for } k \le p \\ -\frac{1}{2k}\sum\frac{1}{z_n^k} & \text{for } k \ge p+1. \end{cases}$$

Recall the notation from Chapter 13; near $z = 0$

$$\log f(z) = \sum_{k=0}^{\infty}\alpha_k z^k.$$

Using the formulas in (13.4.18), we therefore get

(i) $c_0 = \displaystyle\sum_{|z_n|\le r}\log\frac{r}{r_n}$

(ii) $c_m = \dfrac{1}{2m}\displaystyle\sum_{|z_n|\le r}\left(\frac{r}{z_n}\right)^m - \frac{1}{2m}\sum_{|z_n|\le r}\left(\frac{\bar z_n}{r}\right)^m$ if $m \le p$

(iii) $c_m = \dfrac{-r^m}{2m}\displaystyle\sum_{|z_n|>r}\left(\frac{1}{z_n}\right)^m - \frac{1}{2m}\sum_{|z_n|\le r}\left(\frac{\bar z_n}{r}\right)^m$ if $m \ge p+1$.

Choose $\rho' > \rho$. We must prove that for some $M = M(\rho')$

$$|c_m(r)| \le M\frac{r^{\rho'}}{|m|+1} \quad \text{for } m = 0, 1, 2, \ldots .$$

But this estimate is easily derived, as in Chapter 13, from the fact that $n(r) = O(r^{\rho'})$, once we know that the c_m are given by formulas (i), (ii), and (iii). $\square$

The next theorem follows easily from the Miles-Rubel-Taylor Theorem of Chapter 14 but is included here due to its simple deduction from Theorem 15.5.

Theorem. *If f is a meromorphic function in the plane with $\rho = \rho(f) < \infty$, then there exist entire functions g and h, with $\rho(g) \le \rho$ and $\rho(h) \le \rho$, such that $f = g/h$.*

Proof. Let $W = \{W_n\}$ be the poles of f, let $\rho_1 = \rho_1(W)$ be the exponent of convergence of W, and let $p = p(W)$ be the genus of W. Since $N(r,f) \le$

$T(r, f) = O(r^{\rho + \epsilon})$ for each $\epsilon > 0$, we have $\rho_1 \leq \rho$. By Theorem 15.5, $\rho(P_W) = \rho_1$. We now let $g = f \cdot P_W$ and observe that g is entire. Since $\rho(g) \leq \max\{\rho(P_W), \rho(f)\} = \rho$, it follows that $f = g/P_W$ is the desired representation.

It was proved by similar means by Rubel and Taylor that if $\lambda(2r)/\lambda(r) = O(1)$, then every meromorphic function of finite λ-type is the quotient of two entire functions of finite λ-type. However, the proof is long and is subsumed in Miles' proof of the last chapter, so we omit it.

Laguerre's Theorem on Separation Zeros. *If f is a nonconstant entire function with only real zeros, has genus 0 or 1, and is real on the real axis, then the zeros of f' are real and are separated by the zeros of f, and the zeros of f are separated by the zeros of f'.*

Proof. We have either

$$f(z) = Cz^K e^{az}\Pi\left(1 - \frac{z}{z_n}\right)$$

or

$$f(z) = Cz^K e^{az}\Pi\left(1 - \frac{z}{z_n}\right)e^{z/z_n},$$

where the z_n are real, and c and a are real (possibly $a = 0$).

It follows that either

$$\frac{f'(z)}{f(z)} = \frac{k}{z} + a + \sum \frac{1}{z - z_n}$$

or

$$\frac{f'(z)}{f(z)} = \frac{k}{z} + a + \sum \frac{z}{z_n(z - z_n)}.$$

On writing $z = x + iy$ and recalling that a is real, we get, in both cases,

$$Im\frac{f'(z)}{f(z)} = -y\left\{\frac{k}{x^2 + y^2} + \sum \frac{1}{(x - z_n)^2 + y^2}\right\},$$

which does not vanish except for $y = 0$, so that the zeros of f' are real. Since f is real for real z, the theorems of calculus apply. By Rolle's Theorem, there is a zero of f' between two consecutive zeros of f, so that the zeros of f are certainly separated by the zeros of f'. To see that the zeros of f' are separated by the zeros of f, note that

$$\left(\frac{f'}{f}\right)(x) = -\frac{k}{x^2} - \sum \frac{1}{(x - x_n)^2} < 0,$$

so that $\frac{f'(x)}{f(x)}$ is decreasing in any interval free of zeros of f, so that f' cannot have two zeros in any such interval. The case of repeated roots is handled by a suitable convention.

16
Formal Power Series

We consider the formal power series

$$f(z) = a_0 + a_1 z + a_2 z^2 + \ldots,$$

which we usually normalize by $a_0 = 0$.

Let $f_n(z : a) = a + a_1 z + a_2 z^2 + \cdots + a_n z^n$; f_n is a polynomial of "degree" n (possibly $a_n = 0$). We adopt the convention that f_n has n zeros $z_1, z_2, \ldots, z_n$, where, if $a_m \neq 0$ but $a_{m+1} = a_{m+2} = \cdots = a_n = 0$, then $z_{m+1} = z_{m+2} = \cdots = z_n = \infty$. For later use, we shall make the convention $\infty/\infty = 1$.

Let $r_n(a) = \max\{|z| : f_n(z : a) = 0\}$, that is, $r_n(a)$ is the modulus of the largest root of $f_n(z : a) = 0$. Note that if $a_n = 0$, $r_n = \infty$. In a certain sense, $r_n(a)$ measures the disaffinity of f for the value $-a$.

Theorem 16.1. *Given any formal power series f, then*

(16.1)
$$\limsup_{n \to \infty} \frac{r_n(a)}{r_n(b)} \leq 2, \qquad (b \neq 0);$$

if $b = 0$, we have

$$\limsup_{n \to \infty} \frac{r_n(a)}{r_n(0)} \leq 1 + \limsup_{n \to \infty} \left[\frac{1}{r_n(a)} \right]^{\frac{m}{n-m}},$$

where

$$f(z) = a_m z^m + a_{m+1} z^{m+1} + \ldots, a_m \neq 0.$$

Note that if the roles of a and b are interchanged, then (16.1) yields $\frac{1}{2} \leq \liminf_{n \to \infty} \frac{r_n(a)}{r_n(b)}$.

Proof. Let $\alpha_1, \alpha_2, \ldots, \alpha_n$ be the zeros of $f_n(z : a)$ arranged so that $|\alpha_1| \leq |\alpha_2| \leq \cdots \leq |\alpha_n|$. Let $\beta_1, \beta_2, \ldots, \beta_n$ be the zeros of $f_n(z : b)$ arranged so that $|\beta_1| \leq |\beta_2| \leq \cdots \leq |\beta_n|$. Thus $f_n(z : a) = a_n \Pi(z - \alpha_k)$ and $f_n(z : b) = \alpha_n \Pi(z - \beta_k)$, where we shall take $a_n \neq 0$.

Now,

$$(16.2) \qquad f_n(z : b) - f_n(z : a) = b - a = a_n[\Pi(z - \beta_k) - \Pi(z - \alpha_k)].$$

Suppose there exists a sequence of n for which $\{\alpha_n\}$, $\{\beta_n\}$ is such that $|\alpha_n| = \lambda_n|\beta_n|$ with $\lambda_n \geq \lambda > 1$. Otherwise, the conclusion of the theorem is clearly true. Therefore, $r_n(a) = \lambda_n r_n(b)$. Now,

$$|\alpha_n - \beta_k| \geq |\alpha_n| - |\beta_k| = \lambda_n|\beta_n| - |\beta_k| \geq (\lambda_n - 1)|\beta_k|.$$

Letting $z = \alpha_n$ in (16.2) gives $b - a = a_n \Pi(\alpha_n - \beta_k)$. Hence,

$$|b - a| \geq |a_n||\Pi(\alpha_n - \beta_k)| \geq |a_n||\Pi(\lambda_n - 1)|\beta_k|.$$

Dividing through by b and taking nth roots gives

$$\left|1 - \frac{a}{b}\right|^{\frac{1}{n}} \geq \left[\frac{|a_n|}{|b|}\Pi(\lambda_n - 1)|\beta_k|\right]^{\frac{1}{n}}.$$

Now suppose we had $\lambda_n - 1 > 1$. Then

$$\left|1 - \frac{a}{b}\right|^{\frac{1}{n}} \geq \left[\frac{|a_n|}{|b|}(1 + \epsilon)^n \Pi|\beta_k|\right]^{\frac{1}{n}}.$$

But $\Pi|\beta_k| = \frac{|b|}{|a_n|}$, so that supposing $\lambda_n - 1 > 1$ implies $\left|1 - \frac{a}{b}\right|^{\frac{1}{n}} \geq (1 + \epsilon)$; a contradiction. Hence, $\limsup \lambda_n \leq 2$, so that we have $\limsup\limits_{n \to \infty} \frac{r_n(a)}{r_n(b)} \leq 2$.

Suppose now that $b = 0$. Write $f(z) = a_m z^m + a_{m+1} z^{m+1} + \ldots, a_m \neq 0$. We now have $f_n(z : b) = z^m a_n \prod\limits_{k=1}^{n-m} (z - \beta_k)$. Thus, $f_n(z : b) - f_n(z : a) =$

$$-a = a_n[z^m \prod_{k=1}^{n-m} (z - \beta_k) - \prod_{k=1}^{n} (z - \alpha_k)].$$

We again have $|\alpha_n - \beta_k| = (\lambda_n - 1)|\beta_k|$ and $|-a| = |a_n \alpha_n^m \prod\limits_{k=1}^{n-m} (\alpha_n - \beta_k)| =$

$$|a_n|[r_n(a)]^m \prod_{k=1}^{n-m} (\lambda_n - 1)|\beta_k|:$$

$$(16.3) \qquad |a|^{\frac{1}{n-m}} = \left[|a_n|[r_n(a)]^m \prod_{k=1}^{n-m} (\lambda_n - 1)|\beta_k|\right]^{\frac{1}{n-m}}.$$

Now suppose for contradiction that

$$\lambda_n - 1 > |a|^{(\frac{1}{n-m})^2} \left(\frac{1}{|a_m|}\right)^{n-m} \left(\frac{1}{r_n(a)}\right)^{\frac{m}{n-m}}.$$

Then there exist $\{\epsilon_n\}$ such that $\epsilon_n > 0$ and

$$\lambda_n - 1 = |a|^{(\frac{1}{n-m})^2} \left(\frac{1}{|a_m|}\right)^{n-m} \left(\frac{1}{r_n(a)}\right)^{\frac{m}{n-m}} (1 + \epsilon_n).$$

Using this expression in (16.3) and that $\prod_{k=1}^{n-m} |\beta_k| = \left|\frac{a_m}{a_n}\right|$ gives

$$|a|^{\frac{1}{n-m}} \geq |a|^{\frac{1}{n-m}} (1 + \epsilon_n)^{n-m};$$

a contradiction.

Hence,

$$\lambda_n - 1 \leq |a|^{(\frac{1}{n-m})^2} \left(\frac{1}{|a_m|}\right)^{n-m} \left(\frac{1}{r_n(a)}\right)^{\frac{m}{n-m}}.$$

Consequently,

$$\limsup_{n\to\infty} \frac{r_n(a)}{r_n(0)} \leq 1 + \limsup_{n\to\infty} \left(\frac{1}{r_n(a)}\right)^{\frac{m}{n-m}},$$

which completes the proof. Theorem 16.1 has the following corollary.

Corollary. *If f is holomorphic in a neighborhood of 0 (that is, f has a positive radius of convergence), then*

$$\limsup_{n\to\infty} \frac{r_n(a)}{r_n(b)} \leq 2$$

for all b.

Definition. $T_n(f) = -\log r_n(1)$ is called the *discrete characteristic* of f.

Theorem (First Fundamental Theorem) **16.2.** $T_n(f) = T_n(f - a) + O(1)$ *for all a with at most one exception.*

The proof is immediate from the definition of $T_n(f)$ and Theorem 16.1. Now let us examine the case of our exception more closely,

$$\limsup_{n\to\infty} \frac{r_n(1)}{r_n(0)} \leq 1 + \limsup_{n\to\infty} \left(\frac{1}{r_n(1)}\right)^{\frac{m}{n-m}}.$$

Suppose $f_n(z : 1) = 1 + a_1 z + \cdots + a_n z^n$ has zeros $\beta_1, \ldots, \beta_n$ arranged in order of increasing moduli. Thus, $a_n \Pi(-\beta_k) = 1$ and hence $\Pi|\beta_k| = \frac{1}{|a_n|}$. Accordingly, $[r_n(1)]^n \geq \frac{1}{|a_n|}$, and therefore

$$[r_n(1)]^{\frac{m}{n-m}} \geq \left(\frac{1}{|a_n|}\right)^{\frac{1}{n}\frac{m}{n-m}} \quad \text{or} \quad \left(\frac{1}{r_n(1)}\right)^{\frac{m}{n-m}} \leq \left(|a_n|^{\frac{1}{n}}\right)^{\frac{m}{n-m}},$$

which gives us the next result.

Corollary. *If f is holomorphic, then $T_n(f) = T_n(f - a) + O(1)$ for all a.*

Theorem (Kakeya) **16.3.** *Let f be a formal power series not identically zero and $R(f)$ its radius of convergence. Then $R(f) \leq \liminf\limits_{n \to \infty} r_n \leq 2R(f)$.*

Proof. Choose $R' < R(f)$. The f_n have only a finite number of zeros in the disk $D_{R'}$ and $\{f_n(z)\} \to \{f(z)\}$ uniformly in $D_{R'}$. Hence, past a certain n_0 we have, by Hurwitz's Theorem, that the functions f_n have the same number of zeros in $D_{R'}$ as f does. Therefore, the largest zero of f_n must leave $D_{R'}$. Thus $R(f) \leq \liminf\limits_{n \to \infty} r_n$, as desired.

Alternatively, we have seen that $r_n \geq |a_n|^{-\frac{1}{n}}$ and hence $\liminf\limits_{n \to \infty} r_n \geq R(f)$.

To prove the second inequality in the theorem, we use the following result.

Lemma. *Let $P(z) = b_0 + b_1 z + \cdots + b_n z^n = 0$. Then*

$$|z| \leq 2 \max \left(\left| \frac{b_{n-1}}{b_n} \right|, \left| \frac{b_{n-2}}{b_n} \right|^{\frac{1}{2}}, \ldots, \left| \frac{b_0}{b_n} \right|^{\frac{1}{n}} \right).$$

Proof of Lemma. Clearly, $|b_n||z|^n \leq |b_0| + |b_0| + |b_1||z| + \cdots + |b_{n-1}||z|^{n-1}$. Writing $|z| = r$, we have $r^n \leq \left| \frac{b_0}{b_n} \right| + \left| \frac{b_1}{b_n} \right| r + \cdots + \left| \frac{b_0}{b_n} \right| r^{n-1}$. Suppose now that $r > 2 \max \left(\left| \frac{b_{n-1}}{b_n} \right|, \left| \frac{b_{n-2}}{b_n} \right|^{\frac{1}{2}}, \ldots, \left| \frac{b_0}{b_n} \right|^{\frac{1}{n}} \right)$. Then $r^n \leq 2^{-n} r^n + 2^{-(n-1)} r^n + \cdots + 2^{-1} r^n = r^n (\frac{1}{2} + \frac{1}{4} + \cdots + \frac{1}{2^n}) < r^n$, which is impossible and the lemma is proved.

Now let $b_k = a_k R^k$, $P(z) = f_n(Rz) = a_0 + a_1 Rz + \cdots + a_n R^n z^n$. Then the roots of $f_n(Rz) = 0$ are $\frac{1}{R}$ times the roots of $f_n(z) = 0$. By the lemma,

$$\frac{r_n}{R} \leq 2 \max \left(\left| \frac{a_{n-1} R^{n-1}}{a_n R^n} \right|, \left| \frac{a_{n-2} R^{n-2}}{a_n R^n} \right|^{\frac{1}{2}}, \ldots, \left| \frac{a_0}{a_n R^n} \right|^{\frac{1}{n}} \right).$$

Choose $R > R(f)$; it follows that $\{|a_n|R^n\}$ is unbounded. Therefore, for a suitable subsequence, $|a_n|R^n \geq |a_k|R^k$ for all $k \leq n$. In that subsequence, $\frac{r_n}{R} \leq 2$. Hence, $\liminf\limits_{n \to \infty} r_n \leq 2R$. Therefore, $\liminf\limits_{n \to \infty} r_n \leq 2R(f)$, as was to be proved.

Corollary (Okada [28]). *A function f is entire if and only if $\lim\limits_{n \to \infty} T_n(f) = -\infty$.*

Theorem (Tsuji [45]) **16.4.** *Let* $\sigma = \sigma(f) = \limsup\limits_{n\to\infty} \dfrac{\log n}{T_n} = \limsup\limits_{n\to\infty} \dfrac{\log n}{\log r_n}$.
Then $\sigma(f) = \rho(f)$.

Proof. We know that $R(f) = \limsup\limits_{n\to\infty} \dfrac{n\log n}{\log \frac{1}{|a_n|}}$ from Proposition 11.4. Take

$\rho' > \rho$; then, for large n, $\rho' > \dfrac{n\log n}{\log \frac{1}{|a_n|}}$. Hence, $|a_n| \le \dfrac{1}{n^{\frac{n}{\rho'}}}$ and thus

$r_n^n \ge \frac{1}{|a_n|} \ge n^{n/\rho'}$, or $r_n \ge n^{\frac{1}{\rho'}}$. Consequently, $\frac{\log n}{\log r_n} \le \rho'$ for large n; hence, $\sigma \le \rho$ as desired.

In the other direction, suppose on the contrary that $\sigma < \rho$. Choose ρ' such that $\sigma < \rho' < \rho$. Then for n large (say $n \ge n_0$), $r^n \ge n^{\frac{1}{\rho'}}$. Choose $M > 1$ so that for $K = 1, 2, \ldots, n_0, |a_K| \le m\dfrac{2^K}{(K!)^{\frac{1}{\rho'}}}$. We prove by induction that for $n \ge n_0$, $|a_n| \le m\dfrac{2^n}{(n!)^{\frac{1}{\rho'}}}$.

First,

$$|a_n| r_n^n \le |a_{n-1}| r_n^{n-1} + \cdots + |a_1| r_n + 1$$

since $f_n(z : 1) = 1 + a_0 z + \cdots + a_n z^n = 0$. Hence,

$$|a_n| \le |a_{n-1}|\frac{1}{r_n} + |a_{n-2}|\frac{1}{r_n^2} + \cdots + |a_1|\frac{1}{r_n^{n-1}} + \frac{1}{r_n^n}$$
$$\le \frac{|a_{n-1}|}{n^{\frac{1}{\rho'}}} + \frac{|a_{n-2}|}{n^{\frac{2}{\rho'}}} + \cdots + \frac{|a_1|}{n^{\frac{n-1}{\rho'}}} + \frac{1}{n^{\frac{n}{\rho'}}}.$$

Therefore,

$$|a_n| \le M\left\{ \frac{2^{n-1}}{n^{\frac{1}{\rho'}}[(n-1)!]^{\frac{1}{\rho'}}} + \frac{2^{n-2}}{n^{\frac{2}{\rho'}}[(n-2)!]^{\frac{1}{\rho'}}} + \cdots + \frac{2'}{n^{\frac{n-1}{\rho'}}(1!)^{\frac{1}{\rho'}}} + \frac{2^0}{n^{\frac{n}{\rho'}}(0!)^{\frac{1}{\rho'}}} \right\}$$
$$\le M\left\{ \frac{2^{n-1}}{(n!)^{\frac{1}{\rho'}}} + \frac{2^{n-2}}{(n!)^{\frac{1}{\rho'}}} + \cdots + \frac{2'}{(n!)^{\frac{1}{\rho'}}} + \frac{2^0}{(n!)^{\frac{1}{\rho'}}} \right\}$$
$$\le \frac{M}{(n!)^{\frac{1}{\rho'}}}\{1 + 2 + 4 + \cdots + 2^{n-1}\} \le M\frac{2^n}{(n!)^{\frac{1}{\rho'}}},$$

as desired. This now leads to a contradiction of the fact that f is of order ρ, for we have $\frac{1}{|a_n|} \ge \frac{1}{M}\frac{(n!)^{\frac{1}{\rho'}}}{2^n}$. Consequently,

$$\log \frac{1}{|a_n|} \ge \text{constant} + \frac{1}{\rho'}\log n! - n\log 2$$

so that

$$\frac{n \log n}{\log \frac{1}{|a_n|}} \le \frac{n \log n}{\text{constant } + \frac{1}{\rho'} \log n! - n \log 2} = \rho' + o(1)$$

since $\log n! \sim n \log n$. This implies that $\rho(f) \le \rho' < \rho = \rho(f)$; a contradiction.

17
Picard's Theorem and the Second Fundamental Theorem

In this section we shall prove Picard's Theorem, state the second fundamental theorem of Nevanlinna (leaving the proof for the next section), and derive some of its consequences.

We shall use two conventions that will greatly simplify our notation. First, we shall always work "modulo $O(1)$". For example, $A = B$ and $A \leq B$ shall mean that $A - B$ is bounded and that $A - B$ is bounded above, respectively. Second, we shall use a notation of Weyl, writing $\parallel$ in front of a statement to mean that the statement holds with the possible exception of a set of finite length.

Picard's Theorem. *If f is a nonconstant meromorphic function on the complex plane, then f does not omit three values on the sphere.*

It may happen that f omits two values. For example, e^z omits 0 and ∞. Our proof of Picard's Theorem is patterned after our proof of the second fundamental theorem of Nevanlinna, and illustrates the main features of the method in a simpler context.

Proof. Without loss of generality, we may assume that f omits 0, 1, and ∞, so that f, $\frac{1}{f}$, and $\frac{1}{f-1}$ are entire. The next lemma is referred to as the *lemma of the logarithmic derivative.*

Lemma. $\parallel T\left(r, \frac{f'}{f}\right) \leq K' \log r + K'' \log^+ T(r, f)$ *for f omitting 0, 1, and* ∞.

From Poisson's formula (7.1),

$$\log f(z) = \frac{1}{2\pi} \int_{-\pi}^{\pi} \log |f(\rho e^{i\varphi})| \frac{\rho e^{i\varphi} + z}{\rho e^{i\varphi} - z} d\varphi; \quad \rho > r = |z|.$$

Since $f \neq 0$ or ∞, we may differentiate with respect to z to get

$$\frac{f'(z)}{f(z)} = \frac{1}{2\pi} \int_{-\pi}^{\pi} \log |f(\rho e^{i\varphi})| \frac{2\rho e^{i\varphi}}{(\rho e^{i\varphi} - z)^2} d\varphi$$

$$\left| \frac{f'(z)}{f(z)} \right| \leq \frac{2\rho}{(\rho - r)^2} \frac{1}{2\pi} \int_{\pi}^{\pi} \log |f(\rho e^{i\varphi})| d\varphi$$

$$\log^+ \left| \frac{f'(z)}{f(z)} \right| \leq \log^+ \rho + 2\log^+ \frac{1}{\rho - r}$$

$$+ \log^+ \frac{1}{2\pi} \int_{\pi}^{\pi} \log |f(\rho e^{i\varphi})| d\varphi \ (\text{mod } O(1)).$$

Recalling the definition of $m(r, f)$, we find

$$\log^+ \left| \frac{f'(z)}{f(z)} \right| \leq \log^+ \rho + 2\log^+ \frac{1}{\rho - r} + \log^+ m(r, f) + \log^+ m\left(r, \frac{1}{f}\right).$$

Since $f \neq 0$ or ∞, we have $m(r, f) = T(r, f)$ and $m\left(r, \frac{1}{f}\right) = T\left(r, \frac{1}{f}\right)$. Moreover, by the first fundamental theorem of Nevanlinna, we have

$$T(r, f) = T\left(r, \frac{1}{f}\right).$$

Hence,

$$\log^+ \left| \frac{f'(z)}{f(z)} \right| \leq \log^+ \rho + 2\log^+ \frac{1}{\rho - r} + 2\log^+ T(\rho, f).$$

Since the right-hand side is independent of the argument of z, it follows that the same estimate will hold for the average,

$$m\left(r, \frac{f'}{f}\right) \leq \log^+ \rho + 2\log^+ \frac{1}{\rho - r} + 2\log^+ T(\rho, f).$$

Now choose $\rho = r + \frac{1}{\log^+ T(r,f)}$ and use the Borel lemma on $2\log^+ T(\rho, f)$ to find,

$$\left\| \ T\left(r, \frac{f'}{f}\right) \leq \log^+ r + 2\log^+ \log^+ T(r, f) + 3\log^+ T(r, f) \right.$$

$$\leq \log^+ r + 4\log^+ T(r, f).$$

This proves the lemma.

Now let

$$F(z) = \frac{1}{f(z)} + \frac{1}{f(z) - 1}.$$

$|F(z)|$ is large whenever f is near 0 or f is near 1. Of course, the set of values of z where these occur is disjoint: hence, $m(r, F) \geq m(r, f) + m\left(r, \frac{1}{f-1}\right)$.
Therefore,

$$(17.1) \qquad T(r, F) \geq T\left(r, \frac{1}{f}\right) + T\left(r, \frac{1}{f-1}\right) \quad \text{since } f \neq 0, 1.$$

Notice that

$$F(z) = \frac{1}{f(z)} \frac{f(z)}{f'(z)} \left[\frac{f'(z)}{f(z)} + \frac{f'(z)}{f(z) - 1}\right].$$

Thus,

$$(17.2) \quad T(r, F) \leq T\left(r, \frac{1}{f}\right) + T\left(r, \frac{f}{f'}\right) + T\left(r, \frac{f'}{f}\right) + T\left(r, \frac{f'}{f-1}\right).$$

From the first fundamental theorem we have

$$T\left(r, \frac{1}{f}\right) = T(r, f) = T(r, f - 1) = T\left(r, \frac{1}{f-1}\right)$$

and

$$T\left(r, \frac{f}{f'}\right) = T\left(r, \frac{f'}{f}\right).$$

Therefore, on combining (17.1) and (17.2) we obtain

$$2T(r, f) \leq T(r, f) + 2T\left(r, \frac{f'}{f}\right) + T\left(r, \frac{f'}{f-1}\right).$$

Consequently,

$$(17.3) \qquad T(r, f) \leq 2T\left(r, \frac{f'}{f}\right) + T\left(r, \frac{f'}{f-1}\right).$$

The lemma above shows that

$$T\left(r, \frac{f'}{f}\right) \leq k' \log r + k'' \log T(r, f).$$

Applying this lemma to f and $f - 1$ in (17.3) gives

$$T(r, f) \leq k' \log r + k''' \log T(r, f).$$

Of course f is not a rational function since it omits three values. Consequently, by Theorem 10.2,

$$\frac{T(r, f)}{\log r} \to \infty \text{ as } r \to \infty.$$

But we have

$$\left\| \frac{T(r, f)}{\log r} \leq k' + k''' \frac{\log^+ T(r, f)}{\log r}, \right.$$

which is impossible, and the proof by contradiction is complete.

Theorem. *(Second Fundamental Theorem of Nevanlinna). Suppose $f(z)$ is a nonconstant meromorphic function in the plane (or in $|z| < R$, $R < \infty$). Let $a_1, a_2, \ldots, a_q$ be q distinct finite complex numbers. Then*

$$(17.4) \qquad m(r, f) + \sum_{\nu=1}^{q} m\left(r, \frac{1}{f - a_\nu}\right) \leq 2T(r, f) - N(r, f) + s(r),$$

where

$$N_1(r, f) = N\left(r, \frac{1}{f'}\right) + 2N(r, f) - N(r, f')$$

and

$$S(r) = m\left(r, \frac{f'}{f}\right) + m\left(r, \sum_{\nu=1}^{q} \frac{f'}{f - a_\nu}\right).$$

Interpretation. It will be shown that

$$\| \; S(r) = O(\log T(r, f)) + O(\log r).$$

Accordingly, we may think of "S" as standing for "small." To interpret the function $N_1(r, f)$, suppose $f(z) = (z - z_0)^K g(z)$, $K \in \mathbb{Z}$, $g(z) \neq 0$ in some neighborhood of z_0. We analyze the contribution to N_1 due to the behavior of f at z_0 by examining the counting function $n_1(r, f) = n\left(r, \frac{1}{f'}\right) + 2n(r, f) - n(r, f')$. If $K > 0$, then f vanishes at z_0 to order K. Consequently, $n(r, f)$ and $n(r, f')$ are unaffected by the behavior of f near z_0. The contribution to n_1 near $z \cdot$ equals the contribution to $n\left(r, \frac{1}{f}\right)$. Of course, $\frac{1}{f'}$ has a pole of order $K - 1$ at z_0. If $K < 0$, then f has a pole of order $-K$ at z_0 and a similar analysis reveals that n_1 "counts" this pole $K - 1$ times. The case $K = 0$ is trivial. In general we find that poles of order K and zeros of order m are "counted" by n_1, respectively, $K - 1$ and $m - 1$ times.

    The gist of the second fundamental theorem is that for most values of a, the contribution of $m\left(r, \frac{1}{f-a}\right)$ to $T\left(r, \frac{1}{f-a}\right)$ is much smaller than the contribution of $N\left(r, \frac{1}{f-a}\right)$ to it. Thus, for most values of a, $N\left(r, \frac{1}{f-1}\right)$ makes the preponderant contribution to $T\left(r, \frac{1}{f-a}\right)$. Recall that $T\left(r, \frac{1}{f-a}\right) - T(r, f) = O(1)$ by the first fundamental theorem.

Remarks. The following weaker form of the theorem also gives us the Picard Theorem:

$$(17.5) \qquad m(r, f) + \sum_{\nu=1}^{q} m\left(r, \frac{1}{f - a_\nu}\right) \leq 2T(r, f) + S(r).$$

For suppose f omits three values, say $0, 1$, and ∞. Then, since f is entire, $m(r, f) = T(r, f)$ and, since f omits 0 and 1, we have

$$m\left(r, \frac{1}{f}\right) = T\left(r, \frac{1}{f}\right) = T(r, f)$$

and

$$m\left(r, \frac{1}{f-1}\right) = T\left(r, \frac{1}{f-1}\right) = T(r, f-1) = T(r, f).$$

Thus (17.5) implies that $3T(r, f) \leq 2T(r, f) + S(r)$ and hence $T(r, f) \leq S(r)$.

By a lemma to be proved in the next section we assert that

$$\left\| m\left(r, \frac{f'}{f}\right) \leq 6\log^+ T(r, f) + 4\log^+ r. \right.$$

Hence,

$$\left\| S(r) \leq 6(q+1)\log^+ T(r, f) + 4(q+1)\log^+ r \right.$$

and thus

$$T(r, f) \leq S(r) \text{ implies that } \lim_{r \to \infty} \frac{T(r, f)}{\log r} < \infty,$$

which is only true, by Theorem 10.2, for rational functions. Since $f \neq \infty$, f is a polynomial and therefore does not omit three values; a contradiction.

Consequences

Definition. Let $\bar{n}(t, f)$ be the number of poles of f in the closed disk $\bar{\mathbb{D}}_t = \{z \in \mathbb{C} : |z| \leq t\}$ counted once, no matter what the multiplicity. Let $\bar{N}(r, f) = \int_{0+}^r \frac{\bar{n}(t,f) - \bar{n}(o,f)}{t} dt$.

Definition. $\delta(a) = \varliminf_{r \to \infty} \dfrac{m\left(r, \frac{1}{f-a}\right)}{T(r, f)} = 1 - \varlimsup_{r \to \infty} \dfrac{N\left(r, \frac{1}{f-a}\right)}{T(r, f)}.$

$\delta(a)$ is called the *deficiency or defect* of the function for the value a.

Definition. $\theta(a) = \varlimsup_{r \to \infty} \left[\dfrac{N\left(r, \frac{1}{f-a}\right) - \bar{N}\left(r, \frac{1}{f-a}\right)}{T(r, f)}\right].$

Definition. The *branching index* is defined to be $\theta^*(a)$ where

$$\theta^*(a) = 1 - \varlimsup_{r \to \infty} \frac{\bar{N}\left(r, \frac{1}{f-a}\right)}{T(r, f)} = \varliminf_{r \to \infty} \left[1 - \frac{\bar{N}\left(r, \frac{1}{f-a}\right)}{T(r, f)}\right].$$

What contributes to $\theta^*(a)$ is a-points that are taken on by f with multiplicity greater than or equal to 2. So a value a that is taken on often with high multiplicity will have a large branching index.

Remark. $\theta^*(a) \geq \theta(a) + \delta(a)$. Then,

$$\theta^*(a) = \varlimsup_{r\to\infty}\left[1 - \frac{\bar{N}}{T}\right] = \varlimsup_{r\to\infty}\left[\frac{N - \bar{N} + 1 - N}{T}\right]$$

$$\geq \varlimsup_{r\to\infty}\left[\frac{N - \bar{N}}{T}\right] + \varlimsup_{r\to\infty}\left[\frac{1 - N}{T}\right]$$

$$= \theta(a) + \delta(a).$$

The next result follows from the second fundamental theorem.

Theorem. *Let $\{a_\nu\}$ be any finite collection of distinct complex numbers possibly including ∞. Then,*

(17.6)
$$\sum_\nu \theta^*(a_\nu) \leq 2.$$

Proof. First suppose f is not a rational function. From the second fundamental theorem we have $m(r, f) + \sum_{\nu=1}^{q} m\left(r, \frac{1}{f-a_\nu}\right) \leq 2T(r, f) - N_1(r, f) + S(r)$. Adding $N(r, f) + \sum_{\nu=1}^{q} N\left(r, \frac{1}{f-a_\nu}\right)$ to both sides, we get:

$$(q+1)T(r, f) \leq 2T(r, f) + N(r, f) + \sum_{\nu=1}^{q} N\left(r, \frac{1}{f-a_\nu}\right) - N_1(r, f) + S(r).$$

Hence,

$$(q-1)T(r, f) \leq \sum_{\nu=1}^{q} N\left(r, \frac{1}{f-a_\nu}\right) - N\left(r, \frac{1}{f'}\right) + N(r, f') - N(r, f) + S(r).$$

Wherever f takes the value a_ν with multiplicity k, $n\left(r, \frac{1}{f-a_\nu}\right) - n\left(r, \frac{1}{f'}\right)$ counts 1. Also, at a pole of f, $n(r, f') - n(r, f)$ counts 1. Therefore, we may write

$$(q-1)T(r, f) \leq \sum_{\nu=1}^{q} \bar{N}\left(r, \frac{1}{f-a_\nu}\right) + \bar{N}(r, f) + S(r).$$

Hence,

(17.7)
$$(q-1) \leq \sum_{\nu=1}^{q} \frac{\bar{N}\left(r, \frac{1}{f-a_\nu}\right)}{T(r, f)} + \frac{\bar{N}(r, f)}{T(r, f)} + \frac{S(r)}{T(r, f)}.$$

Now f is not a rational function, and therefore $\varliminf_{r\to\infty} \frac{S(r)}{T(r,f)} = 0$ since $\|S(r) = O(\log T(r,f)) + O(\log r)$ by Lemma 18.3 in the next section.

Now, using the fact that $\varlimsup (A+B) \le \varlimsup A + \varlimsup B$, we have

$$(q-1) \le \sum_{\nu=1}^{q} \varlimsup_{r\to\infty} \frac{\bar{N}(r,\frac{1}{f-a_\nu})}{T(r,f)} + \varlimsup_{r\to\infty} \frac{\bar{N}(r,f)}{T(r,f)},$$

from which

$$-\sum_{\nu=1}^{q} \varlimsup \frac{N\left(r,\frac{1}{f-a_\nu}\right)}{T(r,f)} - \varlimsup \frac{\bar{N}(r,f)}{T(r,f)} \le 1 - q.$$

Adding $q+1$ to both sides gives

$$\sum_{\nu=1}^{q} \left[1 - \varlimsup \frac{\bar{N}(r,\frac{1}{f-a_\nu})}{T(r,f)}\right] + 1 - \varlimsup \frac{\bar{N}(r,f)}{T(r,p)} \le 2.$$

Hence, $\sum_{\nu=1}^{q} \theta^*(a_\nu) \le 2$ holds if f is not a rational function.

Now if f is a rational function, the same claims hold up to (17.7). We may suppose $|f(z)| \sim |z|^p$, where p is a nonzero integer. Then $T(r,f) = |p| \log^+ r$. Also,

$$S(r) = m\left(r,\frac{f'}{f}\right) + m\left(r, \sum_{\nu=1}^{q} \frac{f'}{f-a_\nu}\right) = m\left(r,\frac{p}{z}\right) + m\left(r, \sum_{\nu=1}^{q} \frac{pz^{p-1}}{z^p - a_\nu}\right).$$

Now $m\left(r,\frac{p}{z}\right) = \frac{1}{2\pi} \int_{-\pi}^{\pi} \log^+ \left|\frac{p}{z}\right| d\theta$ and, for $|z|$ sufficiently large, $m\left(r,\frac{p}{z}\right) = 0$. Similarly, for $|z|$ large, we have $m\left(r, \sum_{\nu=1}^{q} \frac{pz^{p-1}}{z^p - a_\nu}\right) = 0$. Therefore, it follows that $\varlimsup_{r\to\infty} \frac{S(r)}{T(r,f))} = 0$, and the rest of the proof follows as before.

Corollary. *Let $\{a_\nu\}$ be any finite collection of complex numbers possibly including ∞. Then,*

$$\sum_{\nu=1}^{q} \delta(a_\nu) \le 2.$$

Definition. a is called a *deficient value* of f if $\delta(a) > 0$.

If f never takes on the value a, then $\delta(a) = 1$. The same is true if f takes on the value a very infrequently. In general, $\delta(a)$ measures the tendency of f to omit the value a and $1 - \delta(a)$ measures the tendency of f to take on the value a.

Corollary. *There are at most countably many deficient values.*

Corollary. *There are at most two values for which $\delta > \frac{2}{3}$.*

Definition. *a is said to be a fully branched value of f if f takes the value a with multiplicity either zero or ≥ 2.*

Remark 17.1. If f is entire, f can have at most two fully branched values and this is the best possible, since, for example, ± 1 are fully branched values of $\sin z$. Also, if f is entire, then there are at most two values of a for which $\delta(a) > \frac{1}{2}$.

The reader will find it instructive to work out $\delta(a)$, $\theta(a)$, and $\theta^*(a)$ for some specific functions, such as e^z, $\tan z$, and the Weierstrass $\mathfrak{p}$ function.

Remark. If a is fully branched, then

$$\theta^*(a) = \varliminf\left[1 - \frac{\bar{N}\left(r, \frac{1}{f-a}\right)}{T(r, f)}\right] \geq \varliminf\left[1 - \frac{\bar{N}\left(r, \frac{1}{f-a}\right)}{N\left(r, \frac{1}{f-a}\right)}\right] \quad \text{since } T \geq N.$$

Also, if a is fully branched, then $\dfrac{\bar{N}(r, \frac{1}{f-a})}{N(r, \frac{1}{f-a})} \leq \frac{1}{2}$, and therefore

$\theta^*(a) \geq \frac{1}{2}$. In general, there are at most four fully branched values for a meromorphic function.

The next result shows that Nevanlinna theory may be used to study certain types of exponential identities. As an example, consider the functional equation

$$e^f + e^g = e^h,$$

where f, g, and h are entire functions. Do there exist f, g, and h so that the identity holds? What is the relationship between f, g, and h? We may write

$$e^{f-h} + e^{g-h} = 1.$$

Then the entire function e^{f-h} omits 0. But e^{g-h} omits 0 as well, therefore e^{f-h} must also omit 1 for the identity to hold. By Picard's Theorem, e^{f-h} must be a constant so $f = h+$ const. Similarly, $g = h+$ const.

Lemma (Hiromi-Ozawa [15]). *Let $a_0(z), a,(z), \ldots, a_n(z)$ be meromorphic functions and let $g_1(z), \ldots, g_n(z)$ be entire functions. Further, suppose that*

$$T(r, a_j) = o\left(\sum_{\nu=1}^{n} m(r, e^{g_\nu})\right) \quad j = 0, 1, \ldots, n$$

holds outside a set of finite logarithmic length. If an identity

$$\sum_{\nu=1}^{n} a_\nu(z)e^{g_\nu(z)} = a_0(z)$$

holds, then we have an identity

$$\sum_{\nu=1}^{n} c_\nu a_\nu(z) e^{g_\nu(z)} = 0,$$

where the constants $c_\nu, \nu = 1, \ldots, n$ are not all zero.

Proof. On writing $f' = \frac{f'}{f} f$, we may use the lemma on the logarithmic derivative to estimate $T(r, f')$ when f is entire. For notational simplicity, let $G_\nu(z) = a_\nu(z) e^{g_\nu(z)}$. Then we have

(17.8) $$\sum_{\nu=1}^{n} G_\nu(z) = a_0(z).$$

By differentiating both sides we obtain

(17.9) $$\sum_{\nu=1}^{n} G_\nu^{(\mu)}(z) = a_0^{(\mu)}(z),$$

which may be rewritten as

(17.10) $$\sum_{\nu=1}^{n} G_\nu(z) \frac{g_\nu^{(\mu)}(z)}{g_\nu(z)} = a_0^{(\mu)}(z), \ \mu = 1, \ldots, m-1.$$

We regard this as a system of simultaneous linear equations in the G_ν. Now we have

$$G_\nu^{(\mu)}(z) = P_\mu(a_\nu, a_\nu', \ldots, a_\nu^{(\mu)}, g_\nu', \ldots, g_\nu^{(\mu)}) e^{g_\nu(z)}$$

with a suitable polynomial P_μ in the indicated functions. Thus we have

(17.11) $$T\left(r, \frac{G_\nu^{(\mu)}}{G_\nu}\right) \le O(T(r, a_\nu) + T(r, g_\nu)) = o\left(\sum_{\nu=1}^{n} m(r, e^{g_\nu})\right)$$

outside a set of finite logarithmic length.

Suppose, for the simultaneous equations (17.8) and (17.10), that the determinant $\Delta \ne 0$. By solving (17.10) with respect to $G_j, j = 1, \ldots, n$ we have by Cramers' rule

$$G_j = \frac{\Delta_j}{\Delta},$$

where

$$\Delta = \begin{vmatrix} 1 & \cdots & 1 \\ G_1^1/G_1 & \cdots & G_n^1/G_n \\ \vdots & \ddots & \vdots \\ G_1^{(n-1)}/G_n & \cdots & G_n^{(n-1)}/G_n \end{vmatrix}$$

and

$$\Delta_j = \begin{vmatrix} 1 & \cdots & 1 & a_0 & 1 & \cdots & 1 \\ \dfrac{G_1^1}{G_1} & \cdots & \dfrac{G_{j-1}^1}{G_{j-1}} & a_0^1 & \dfrac{G_{j+1}^1}{G_{j+1}} & \cdots & \dfrac{G_n^1}{G_n} \\ & & & \vdots & & & \\ \dfrac{G_1^{(n-1)}}{G_1} & \cdots & \dfrac{G_{j-1}^{(n-1)}}{G_{j-1}} & a_0^{(n-1)} & \dfrac{G_{j+1}^{(n-1)}}{G_{j+1}} & \cdots & \dfrac{G_n^{(n-1)}}{G_n} \end{vmatrix}.$$

Since

(17.12)
$$T\left(r, \frac{G_\nu^{(\mu)}}{G_\nu}\right) = o\left(\sum_{\nu=1}^n m(r, e^{g_\nu})\right),$$

we have

$$T(r, \Delta) = O\left(\sum_{\nu=1}^n m(r, e^{g_\nu})\right), T(r, \Delta_j) = o\left(\sum m(r, e^{g_\nu})\right)$$

for $j = 1, \ldots, n$ outside a set of finite logarithmic length. Thus we have

$$m(r, e^{g_\nu}) = T(r, e^{g_\nu}) \le T(r, a_\nu) + T(r, G_\nu)$$

$$\le T(r, a_\nu) + T(r, \Delta) + T(r, \Delta_\nu) = o\left(\sum_{\nu=1}^n m(r, e^{g_\nu})\right)$$

and hence

$$\sum_{\nu=1}^n m(r, e_\nu^g) = o\left(\sum_{\nu=1}^n m(r, e^{g_\nu})\right)$$

outside a set of finite Lebesgue measure. But this is a contradiction. Consequently, the Wronskian $\Delta \equiv 0$ and the result follows.

We say that two meromorphic functions f and g *share the value* a ($a = \infty$ is allowed) if $f(z) = a$ whenever $g(z) = a$ and also $g(z) = a$ whenever $f(z) = a$, counting multiplicities in both cases. A famous theorem of R. Nevanlinna, which will be proven shortly, implies that if two nonconstant meromorphic functions f and g on the complex plane share five distinct finite values (ignoring multiplicity), then it follows that $f = g$, and the number 5 cannot be reduced. We consider here the special case $g = f'$, the derivative of f, and prove the following result.

Theorem. *If f is a nonconstant entire function in the finite complex plane, and if f and f' share two distinct finite values (counting multiplicity), then $f' = f$.*

In other words, a derivative is worth two values. We show at the end of the chapter that the number 2 of the theorem cannot be reduced. We do not now know whether there is a result corresponding to our theorem if one ignores multiplicities, or if one considers *meromorphic* instead of *entire* functions.

Proof of Theorem. To fix the ideas, we suppose that f and f' share the values a and b, where $a = 1$ and $b = 2$. Other choices of a and b make no real difference, except if a or b is zero, in which case the analysis becomes easier, and is left to the reader. We may write then

$$(17.13) \qquad \frac{f' - 1}{f - 1} = e^{k_1}$$

$$(17.14) \qquad \frac{f' - 1}{f - 2} = e^{k_2},$$

where k_1 and k_2 are entire functions. We solve (17.8) and (17.9) for f to get

$$(17.15) \qquad f = \frac{1 + e^{k_1} - 2e^{k_2}}{e^{k_1} - e^{k_2}}.$$

We now differentiate both sides of (17.10) and substitute (17.8) to get
(17.16)
$$2e^{2k_1} + e^{2k_2} + (k_2' - k_1' - 3)e^{k_1+k_2} - e^{2k_1+k_2} + e^{k_1+2k_2} + k_1'e^{k_1} - k_2'e^{k_2} = 0.$$

We shall make repeated use of the lemma of Hiromi and Ozawa.
Now, in (17.16), divide by e^{k_2} to get

$$(17.17) \quad 2e^{2k_1-k_2} + e^{k_2} + (k_2' - k_1' - 3)e_1^k - 2^{2k_1} + e^{k_1+k_2} + k_j'r^{k_1-k_2} = k_2'.$$

We now apply the lemma to get
(17.18)
$$c_1e^{2k_1-k_2} + c_2e^{k_2} + c_3(k_2' - k_1' - 3)e^{k_1} + c_4e^{2k_1} + c_5e^{k_1^+k_2} + c_6k_1'e^{k_1-k_2} = 0,$$

where $c_1, \ldots, c_6$ are constants that are not all zero.

The hypotheses of the Hiromi-Ozawa Lemma are satisfied because, for example, k_1' is the logarithmic derivative of e^{k_1} and we may use the lemma of the logarithmic derivative. It follows that $T(r, k') = 0(T(r, e^k))$, outside of a suitably small exceptional set. At any rate, we divide in (17.17) by e^{k_1} to get

$$(17.19) \quad c_1e^{k_1-k_2} + c_2e^{k_2-k_1} + c_4e^{k_1} + c_5e^{k_2} + c_6k_1'e^{-k_2} = c_3(k_2' - k_1' - 3),$$

and we may use the lemma once more to get

$$(17.20) \qquad d_1e^{k_1-k_2} + d_2e^{k_2-k_1} + d_3e^{k_1} + d_4e^{k_2} + d_5k_1'e^{-k_2} = 0$$

for suitable $d_1, \ldots, d_5$. Multiply by e^{k_2} to get

$$(17.21) \qquad d_1 e^{k_1} + d_2 e^{2k_2 - k_1} + d_3 e^{k_1 + k_2} + d_4 e^{2k_2} = -d_5 k_1'$$

and apply the lemma yet again to get

$$(17.22) \qquad u_1 e^{k_1} + u_2 e^{2k_1 - k_2} + u_3 e^{k_1 + k_2} + u_4 e^{2k_2} = 0,$$

where u_1, u_2, u_3, u_4 are constants that are not all zero. Now, by successive applications of the lemma, we reach a contradiction, unless possibly one of the five following conditions holds for some constant C:

$$k_1 = k_2 + C, l_2 = C, k_1 = 2k_2 + C, k_2 = 2k_1 + C, k_1 = C.$$

We now rule out these possibilities unless $f' = f$. First, it is easy to see that $k_1 = C$ (and similarly $k_2 = C$) is consistent with (17.13) and (17.14) unless $d = e^C = 1$, in which case $f' = f$. For if $(f' - 1)/(f - 1) = d$, then $f = (d-1)/d + be^{dz}$ for some constant b, and hence $f' = bde^{dz}$. This clearly contradicts (17.14) unless $d = 1$, for we would have

$$(17.23) \qquad \frac{f' - 2}{f - 2} = \frac{bde^{dz} - 2}{be^{dz} - 2} = e^{k_2},$$

which is impossible unless $d = 1$ (remember that f is not constant, so that $b \neq 0$).

Next, we rule out $k_1 = k_2 + C$. We go back to (17.16) to get

$$(17.24) \qquad m_1 e^{2k_1} + m_2 e^{3k_1} = k_1' m_4 e^{k_1}.$$

Apply the lemma again after dividing by e^{k_1} to get

$$(17.25) \qquad n_1 e^{k_1} + n_2 e^{2k_1} = 0,$$

which implies that $k_1 = 2k_2 + C$ (and similarly, $k_2 = 2k_1 + C$) is impossible unless $f = f'$. From (17.24) we would get

$$(17.26) \qquad d_2 e^{-C} + (d_1 e^C + d_4) e^{2k_2} + d_3 e^C e^{3k_2} = -2d_5 k_2'$$

and apply the lemma for the last time to get

$$(17.27) \qquad \ell_1 + \ell_2 e^{2k_2} + \ell_3 e^{3k_2} = 0,$$

where ℓ_1, ℓ_2, ℓ_3 are constants that are not all zero. In other words, $P(e^{k_2}) = 0$, where P is a cubic polynomial, so e^{k_2} is a constant, which we already have ruled out unless $f = f'$. This completes the proof of the theorem.

Finally, it is easy to see that there exists a nontrivial entire function that does share *one* value with its derivative. For example,

$$f(z) = e^{e^z} \int_0^z e^{-e^t}(1 - e^t)dt$$

satisfies $(f' - 1)/(f - 1) = e^z$ so that f and f' share the value 1. This shows that the number *two* of our theorem is the best possible.

Now we come, as an application of the second fundamental theorem, to a truly beautiful and surprising theorem of Nevanlinna. Recall that, given two meromorphic functions $f_1(z)$ and $f_2(z)$, and a complex number (finite or infinite) w, we say that f_1 and f_2 *share* the value w (ignoring multiplicity) if every z for which $f_1(z) = w$ also satisfies $f_2(z) = w$, and vice versa. We use the same terminology if both functions *omit* the value w.

Theorem. *If two functions, meromorphic in the whole complex plane* $\mathbb{C}$, *share five distinct values, then the two functions must be the same.*

Note that e^z and e^{-z} share $0, \infty, 1, -1$, so the number *five* is sharp.

Proof of Theorem. Given a number w, finite or not, let $n_0(r, w)$ be the number of common roots of $f_1(z) = w$ and $f_2(z) = w$ contained in the disc $\{|z| \leq r\}$, each counted only one time. Then put

$$N_0(r, w) = \int_0^r \frac{n_0(t, w) - n_0(0, w)}{t}\, dt + n_0(0, w) \log r$$

and

$$N_{12}(r, w) = \bar{N}\left(r, \frac{1}{f_1 - w}\right) + \bar{N}\left(r, \frac{1}{f_2 - w}\right) - 2N_0(r, w).$$

Now, taking for $w_1, \ldots, w_q$ distinct finite complex numbers and applying the second fundamental theorem to the functions f_1 and f_2, we have, off a possible exceptional set of finite length,

$$(q - 2)(T(r, f_1) + T(r, f_2)) < \sum_{\nu=1}^q \left(\bar{N}\left(r, \frac{1}{f_1 - w_\nu}\right) + \bar{N}\left(r, \frac{1}{f_2 - w_\nu}\right)\right)$$
$$+ O[\log rT(r, f_1)T(r, f_2))]$$
$$= \sum_{\nu=1}^q N_{12}(r, w_\nu) + 2\sum_{\nu=1}^q N_0(r, w_\nu)$$
$$+ O[\log rT(r, f_1)T(r, f_2)].$$

Now, if the functions f_1 and f_2 are not the same, every common root of the equations $f_1 = w_\nu$, $f_2 = w_\nu$, is a pole of the function $\frac{1}{f_1 - f_2}$. We deduce from this that

$$\sum_{\nu=1}^q N_0(r, w_\nu) \leq N\left(r, \frac{1}{f_1 - f_2}\right) < T(r, f_1 - f_2) + O(1).$$

On the other hand,

$$T(r, f_1 - f_2) < T(r, f_1) + T(r, f_2) + O(1).$$

We conclude

$$(q - 4)(T(r, f_1) + T(r, f_2)) < \sum_{\nu=1}^{q} N_{12}(r, w_\nu)$$
$$+ O[\log(rT(r, f_1)T(r, f_2))]$$

off a set of exceptional segments of finite total length.

(If one of the w_ν is infinite, just apply a linear fractional transformation to f_1 and f_2.)

Suppose to begin with that one of the two functions, say f_1, is transcendental. Then among the five given values $w = w_\nu$, $\nu = 1, 2, \ldots, 5$, there are at least three that are taken by f_1 in an infinite number of points z. By hypothesis, the same is true for f_2. Hence, f_2 is also transcendental. Suppose for a moment that f_1 and f_2 are not the same. Apply the last inequality above, which shows that for $q = 5$, the expression N_{12} vanishing for the five given values,

$$T(r, f_1) + T(r, f_2) < O(\log[rT(r, f_1)T(r, f_2)]).$$

This implies that

$$\lim_{r \to \infty} \frac{T(r, f_1)}{\log r} < \infty \quad \text{and} \quad \lim_{r \to \infty} \frac{T(r, f_2)}{\log r} < \infty.$$

This is impossible since f_1 and f_2 are both not rational functions. The theorem thus is proved by contradiction in the case that one of the given functions is transcendental.

But the conclusion is obvious if both of the functions are rational, since a rational function f is uniquely determined by the roots of $f(z) = w$ for any *three* distinct values of w.

18
A Proof of the Second Fundamental Theorem

We now begin the proof of the second fundamental theorem of Nevanlinna. We continue to use the convention that all equalities are to be read "modulo $O(1)$." There are a number of other proofs of the second fundamental theorem, some of them leading to generalizations of it.

Lemma 18.1. *Let f be a meromorphic function in the plane, and let $a_1, \ldots, a_n$ be distinct complex numbers. Let*

$$F(z) = \sum_{\nu=1}^{n} \frac{1}{f(z) - a_\nu}.$$

Then

$$m(r, F) \geq \sum_{\nu=1}^{n} m\left(r, \frac{1}{f - a_\nu}\right).$$

Proof. We first remark that the lemma is intuitively obvious, for when $f(z)$ is "close to" a_ν it contributes to $m\left(r, \frac{1}{f - a_\nu}\right)$, but not to any $m\left(r, \frac{1}{f - a_j}\right)$ with $j \neq \nu$.

To prove the lemma in detail, we introduce the following notation. Let $\delta > 0$ be given with $2\delta \leq \min\{|a_\nu - a_j| : 1 \leq \nu < j \leq n\}$. We require $\delta \leq 1$. Let

$$E'_\nu = f^{-1}\{w : |w - a_\nu| < \delta\} = \{z : |f(z) - a_\nu| < \delta\}$$
$$E'_\nu(r) = E'_\nu \cap \{z : |z| = r\}$$
$$E_\nu(r) = \{\theta : re^{i\theta} \in E'_\nu(r)\}$$
$$C_\nu(r) = [0, 2\pi] - E_\nu(r).$$

In other words, $E_\nu(r)$ is the set of all θ for which $|f(re^{i\theta}) - a_\nu| < \delta$, i.e., those θ's for which $f(re^{i\theta})$ contributes significantly to $m\left(r, \frac{1}{f-a_\nu}\right)$. It is clear that $E_\nu(r)$ and $E_j(r)$ are disjoint provided that $\nu \neq j$.

Moreover,

$$\frac{1}{2\pi}\int_{-\pi}^{\pi} \log^+ |F(re^{i\theta})| d\theta \geq \frac{1}{2\pi}\int^* \log^+ |F(re^{i\theta})| d\theta,$$

where $\int^*$ is over the set $\bigcup_{\nu=1}^n E_\nu(r)$. Now

$$\frac{1}{2\pi}\int^* = \sum_{\nu=1}^n \frac{1}{2\pi}\int_{E_\nu(r)} \log^+ |F(re^{i\theta})| d\theta$$

$$\geq \sum_{\nu=1}^n \frac{1}{2\pi}\int_{E_\nu(r)} \log^+ \left|\frac{1}{f(re^{i\theta}) - a_\nu}\right| d\theta$$

$$- \frac{1}{2\pi}\int_{E_\nu(r)} \log^+ \left|\sum_{\substack{j=1 \\ j\neq\nu}}^n \frac{1}{f(re^{i\theta}) - a_j}\right| d\theta.$$

But

$$m\left(r, \frac{1}{f - a_\nu}\right) = \frac{1}{2\pi}\int_{-\pi}^{\pi} \log^+ \left|\frac{1}{f(re^{i\theta}) - a_\nu}\right| d\theta$$

$$= \frac{1}{2\pi}\int_{E_\nu(r)} \log^+ \left|\frac{1}{f(re^{i\theta}) - a_\nu}\right| d\theta + \frac{1}{2\pi}\int_{C_\nu(r)} \log^+ \left|\frac{1}{f(re^{i\theta}) - a_\nu}\right| d\theta.$$

$$= \frac{1}{2\pi}\int_{E_\nu(r)} \log^+ \left|\frac{1}{f(re^{i\theta}) - a_\nu}\right| d\theta + O(1).$$

Also,

$$\frac{1}{2\pi}\int_{E_\nu(r)} \log^+ \left|\sum_{\substack{j=1 \\ j\neq\nu}}^n \frac{1}{f(re^{i\theta}) - a_j}\right| \leq \frac{n-1}{\delta} = O(1).$$

Hence we see that

$$m(r, F) \geq \sum_{\nu=1}^n \frac{1}{2\pi}\int_{E_\nu(r)} \log^+ \left|\frac{1}{f(re^{i\theta}) - a_\nu}\right| d\theta$$

$$- \frac{1}{2\pi}\int_{E_\nu(r)} \log^+ \left|\sum_{\substack{j=1 \\ j\neq\nu}}^n \frac{1}{f(re^{i\theta}) - a_j}\right| d\theta$$

$$\geq \sum_{\nu=1}^n m\left(r, \frac{1}{f - a_\nu}\right) + O(1),$$

which proves Lemma 18.1.

Definition. $N_1(r, f) = N\left(r, \frac{1}{f'}\right) + 2N(r, f) - N(r, f')$. $N_1(r, f)$ is interpreted in the previous chapter after the statement of the second fundamental theorem.

Lemma 18.2.

$$\sum_{\nu=1}^{n} m\left(r, \frac{1}{f - a_\nu}\right) \leq 2T(r, f) - N_1(r) - m(r, f)$$

$$+ m\left(r, \frac{f'}{f}\right) + m\left(r, \sum_{\nu=1}^{n} \frac{f'}{f - a_\nu}\right).$$

Proof. Let $F(z)$ be as in Lemma 18.1 and write

$$F(z) = \frac{1}{f(z)} \cdot \frac{f(z)}{f'(z)} \sum_{\nu=1}^{n} \frac{f'(z)}{f(z) - a_\nu}.$$

Then

(18.1) $\qquad m(r, F) \leq m\left(r, \frac{1}{f}\right) + m\left(r, \frac{f}{f'}\right) + m\left(r, \sum_{\nu=1}^{n} \frac{f'}{f - a_\nu}\right).$

But $m\left(r, \frac{1}{f}\right) = T(r, f) - N\left(r, \frac{1}{f}\right)$ and $m\left(r, \frac{f}{f'}\right) = m\left(r, \frac{f'}{f}\right) + N\left(r, \frac{f'}{f}\right) - N\left(r, \frac{f}{f'}\right)$. Hence,

(18.2)
$$m(r, F) \leq T(r, f) - N\left(r, \frac{1}{f}\right) + m\left(r, \frac{f'}{f}\right) + N\left(r, \frac{f'}{f}\right) - N\left(r, \frac{f}{f'}\right)$$

$$+ m\left(r, \sum_{\nu=1}^{n} \frac{f'}{f - a_\nu}\right).$$

Now observe that if we define

$$\varphi(r, g) = N(r, g) - N\left(r, \frac{1}{g}\right),$$

then

$$\varphi(r, gh) = \varphi(r, g) + \varphi(r, h).$$

Using this observation in (18.2) we see that
(18.3)
$$m(r, F) \leq T(r, f) - N\left(r, \frac{1}{f}\right) + m\left(r, \frac{f'}{f}\right) + N(r, f') + N\left(r, \frac{1}{f}\right)$$

$$- N\left(r, \frac{1}{f'}\right) - N(r, f) + m\left(r, \sum_{\nu=1}^{n} \frac{f'}{f - a_\nu}\right).$$

If we now add and subtract $T(r, f) = m(r, f) + N(r, f)$ on the right-hand side, we obtain

(18.4)
$$m(r, F) \le 2T(r, f) - m(r, f) + m\left(r, \frac{f'}{f}\right) + m\left(r, \sum_{\nu=1}^{n} \frac{f'}{f - a_\nu}\right)$$
$$- \left\{2N(r, f) - N(r, f') + N\left(r, \frac{1}{f'}\right)\right\}$$

or

$$m(r, F) \le 2T(r, f) - m(r, f) + m\left(r, \frac{f'}{f}\right) + m\left(r, \sum_{\nu=1}^{n} \frac{f'}{f - a_\nu}\right) - N_1(r).$$

By Lemma 1, $m(r, F) \ge \sum_{\nu=1}^{n} m\left(r, \frac{1}{f - a_\nu}\right)$, and Lemma 18.2 follows.

We are now ready to prove the second fundamental theorem of Nevanlinna.

Theorem. *Let f be a meromorphic function in the plane and let*

$$S(r) = m\left(r, \frac{f'}{f}\right) + m\left(r, \sum_{\nu=1}^{n} \frac{f'}{f - a_\nu}\right).$$

Then

$$m(r, f) + \sum_{\nu=1}^{n} m\left(r, \frac{1}{f - a_\nu}\right) \le 2T(r, f) - N_1(r) + S(r).$$

Proof. This is just Lemma 18.2.

We saw previously that in order to deduce the Picard Theorem from the second fundamental theorem, we needed an estimate on the size of $S(r)$. Such an estimate follows from:

Lemma (Lemma of the Logarithmic Derivative) **18.3.** *If f is meromorphic in the plane and $0 < r < \rho$, then*

$$m\left(r, \frac{f'}{f}\right) \le 4\log^+ \rho + 3\log^+ \frac{1}{\rho - r} + 4\log^+ T(\rho, f).$$

Proof. Without loss of generality, we suppose that $f(0) \ne 0, \infty$ since, if $g(z) = z^k f(z)$,
$$m\left(r, \frac{g'}{g}\right) - m\left(r, \frac{f'}{f}\right) = O(1).$$

Let $z = re^{i\theta}$. Then for a suitably defined branch of the logarithm we have by the Poisson-Jensen formula,

(18.5)
$$
\begin{aligned}
\log f(z) = {} & \frac{1}{2\pi} \int_{-\pi}^{\pi} \log |f(\rho e^{i\varphi})| \frac{\rho e^{i\varphi} + z}{\rho e^{i\varphi} - z} d\varphi \\
& + \sum_{|z_\nu| < \rho} \log B_\rho(z, z_\nu) - \sum_{|W_\nu| < \rho} \log B_\rho(z, w_\nu) + i\lambda,
\end{aligned}
$$

where $B_\rho(z, a)$ is the Blaschke factor mapping the disk of radius ρ onto the unit disk and a onto 0, λ is a real constant, and, as usual, z_ν and w_ν are the zeros and poles of f, respectively. [Equation (18.1) holds without any $O(1)$ terms.] Thus, by differentiating (18.5), we obtain
(18.6)
$$
\begin{aligned}
\frac{f'(z)}{f(z)} = {} & \frac{1}{2\pi} \int_{-\pi}^{\pi} \log |f(\rho e^{i\varphi})| \frac{2\rho e^{i\varphi}}{(\rho e^{i\varphi} - z)^2} d\varphi + \sum_{|z_\nu| \le \rho} \frac{\rho^2 - |z_\nu|^2}{(z - z_\nu)(\rho^2 - \overline{z_\nu} z)} \\
& - \sum_{|w_\nu| \le \rho} \frac{\rho^2 - |w_\nu|^2}{(z - w_\nu)(\rho^2 - \overline{w_\nu} w)}.
\end{aligned}
$$

Taking simple estimates, we have
(18.7)
$$
\left| \frac{f'(z)}{f(z)} \right| \le \frac{2\rho}{(\rho - r)^2} \frac{1}{2\pi} \int_{-\pi}^{\pi} |\log |f(\rho e^{i\varphi})|| d\varphi + \sum_{|\lambda_\nu| \le \rho} \frac{\rho^2 - |\lambda_\nu|^2}{|z - \lambda_\nu||\rho^2 - \bar{\lambda}_\nu z|},
$$

where λ_ν now runs over both the zeros and the poles of f. Observe that

$$
\begin{aligned}
\frac{\rho^2 - |\lambda_\nu|^2}{|z - \lambda_\nu||\rho^2 - \bar{\lambda}_\nu z|} &= \left| \frac{\rho^2 - \bar{\lambda}_\nu z}{\rho(z - \lambda_\nu)} \right| \cdot \frac{\rho(\rho^2 - |\lambda_\nu|^2)}{|\rho^2 - \bar{\lambda}_\nu z|^2} \\
&= \left| \frac{1}{B_\rho(z, \lambda_\nu)} \right| \frac{\rho(\rho^2 - |\lambda_\nu|^2)}{|\rho^2 - \bar{\lambda}_\nu z|^2} \le \left| \frac{1}{B_\rho(z, \lambda_\nu)} \right| \frac{\rho}{(\rho - r)^2}.
\end{aligned}
$$

The last step follows from $|\rho^2 - \bar{\lambda}_\nu z| \ge \rho^2 - |\bar{\lambda}_\nu||z| \ge \rho^2 - \rho r = \rho(\rho - r)$.

If we use this estimate in (18.7), take $\log^+$ of both sides of (18.7), and apply the additive inequality satisfied by $\log^+$, $(\log^+(a, + \cdots + a_n) \le \log^+ a, + \cdots + \log^+ a_n + \log n)$, we then find that
(18.8)
$$
\begin{aligned}
\log^+ \left| \frac{f'(z)}{f(z)} \right| \le {} & \log^+ \rho + 2\log^+ \frac{1}{\rho - r} + \log^+ \frac{1}{2\pi} \int_{-\pi}^{\pi} |\log^+ |f(re^{i\varphi})|| d\varphi \\
& + \sum_{|\lambda_\nu| \le \rho} \log^+ \frac{1}{|B_\rho(z, \lambda_\nu)|} + \log^+ \left(n(\rho, f) + n\left(\rho, \frac{1}{f}\right) \right).
\end{aligned}
$$

Since $|z| = r < \rho$, $\frac{1}{|B_\rho(z,\lambda_\nu)|} \ge 1$, we see that $\log^+ \frac{1}{|B_\rho(z,\lambda_\nu)|} = \log \frac{1}{|B_\rho(z,\lambda_\nu)|}$.

Therefore, by Jensen's Theorem,

(18.9)
$$\frac{1}{2\pi}\int_{-\pi}^{\pi}\log^{+}\left|\frac{1}{B_{\rho}(re^{i\theta},\lambda_{\nu})}\right|\,d\theta = \begin{cases} \log\frac{\rho}{|\lambda_{\nu}|} - \log\frac{r}{|\lambda_{\nu}|} & \text{if } |\lambda_{\nu}| \leq r \\ \log\frac{\rho}{|\lambda_{\nu}|} & \text{if } |\lambda_{\nu}| > r. \end{cases}$$

To simplify the notation, let $n(r) = n(r,f) + n(r,\frac{1}{f})$ and likewise $N(r) = N(r,f) + N(r,\frac{1}{f})$. If we let $z = re^{i\theta}$ and integrate (18.8) over the circle of radius r with center at 0, we obtain

(18.10)
$$m\left(r,\frac{f'}{f}\right) = \frac{1}{2\pi}\int_{-\pi}^{\pi}\log^{+}\left|\frac{f'(re^{i\theta})}{f(re^{i\theta})}\right|\,d\theta$$

$$\leq \log^{+}\rho + 2\log^{+}\frac{1}{\rho-r} + \log^{+}\frac{1}{2\pi}\int_{-\pi}^{\pi}|\log^{+}|f(re^{i\varphi})||\,d\varphi$$

$$+ \sum_{|\lambda_{\nu}|\leq\rho}\frac{1}{2\pi}\int_{-\pi}^{\pi}\left|\log^{+}|\frac{1}{B_{\rho}(re^{i\theta},\lambda_{\nu})}|\right|\,d\theta + \log^{+}n(\rho).$$

We recognize that $\frac{1}{2\pi}\int_{-\pi}^{\pi}|\log^{+}|f(\rho e^{i\theta})||d\varphi = m(\rho,f) + m\left(\rho,\frac{1}{f}\right) = O(2T(\rho,f))$ and

$$\sum_{|\lambda_{\nu}|\leq\rho}\frac{1}{2\pi}\int_{-\pi}^{\pi}\log^{+}\left|\frac{1}{B_{\rho}(re^{i\varphi},\lambda_{\nu})}\right|\,d\varphi$$

$$= \sum_{|\lambda_{\nu}|\leq\rho}\log\left(\frac{\rho}{|\lambda_{\nu}|}\right) - \sum_{|\lambda_{\nu}|\leq r}\log\left(\frac{r}{|\lambda_{\nu}|}\right) = N(\rho) - N(r).$$

Hence,

(18.11)
$$m\left(r,\frac{f'}{f}\right) \leq \log^{+}\rho + 2\log^{+}\frac{1}{\rho-r} + \log^{+}T(\rho,f) + \log^{+}n(\rho) + N(\rho) - N(r).$$

We now will estimate the term $\log^{+}n(\rho)$. Choose a number ρ' with $\rho' > r$. Then

$$n(\rho) = \frac{n(\rho)}{\log(\frac{\rho'}{\rho})}\int_{\rho}^{\rho'}\frac{dt}{t} \leq \frac{1}{\log(\frac{\rho'}{\rho})}\int_{\rho}^{\rho'}\frac{n(t)}{t}\,dt$$

$$\leq \frac{N(\rho')}{\log(\frac{\rho'}{\rho})} \leq \frac{2T(\rho',f)+C}{\log(\frac{\rho'}{\rho})},$$

where C is an appropriate constant. Now

$$\frac{1}{\rho'}(\rho'-\rho) \leq \int_{\rho}^{\rho'}\frac{dt}{t} = \log\frac{\rho'}{\rho} \leq \frac{1}{\rho}(\rho'-\rho).$$

Hence,

$$n(\rho) \leq \frac{2T(\rho',f)+C}{\frac{1}{\rho'}(\rho'-\rho)}$$

so that

$$\log^+ n(\rho) \leq \log^+ \rho' + \log^+ \frac{1}{\rho'-\rho} + \log^+ T(\rho',f).$$

Hence,

$$m\left(r,\frac{f'}{f}\right) \leq \log^+ \rho + \log^+ \rho' + 2\log^+ \frac{1}{\rho-r} + \log^+ \frac{1}{\rho'-\rho} + \log^+ T(\rho,f)$$
$$+ \log^+ T(\rho',f) + N(\rho) - N(r).$$

Therefore, with $r < \rho < \rho'$ we have
(18.12)
$$m\left(r,\frac{f'}{f}\right) \leq 2\log^+ \rho' + 2\log^+ \frac{1}{\rho-r} + \log^+ \frac{1}{\rho'-\rho} + 2\log^+ T(\rho',f)$$
$$+ N(\rho) - N(r).$$

We now want to make the term $N(\rho) - N(r)$ small. To do this we use the logarithmic convexity of N. [N is logarithmically convex, since $n(t)$ is increasing and $N(r) = \int_0^r \frac{n(t)}{t}dt$]. Since $r < \rho < \rho'$ and N is logarithmically convex, we have

$$N(\rho) - N(r) \leq \frac{\log(\frac{\rho}{r})}{\log(\frac{\rho'}{r})}(N(\rho') - N(r))$$

$$\leq \frac{\frac{1}{r}(\rho-r)}{\frac{1}{\rho'}(\rho'-r)}(N(\rho') - N(r)) \leq \frac{\rho'(\rho-r)}{r(\rho'-r)}N(\rho')$$

$$\leq \frac{\rho'(\rho-r)}{r(\rho'-r)}[2T(\rho',f)+C],$$

where C is an appropriate constant, $C > 1$.

Now, considering $\frac{\rho'}{r}\left(\frac{\rho-r}{\rho'-r}\right)[2T(\rho',f)+C]$ as a function of ρ, we see that it vanishes for $\rho = r$ and is greater than 1 for $\rho = \rho'$. Since it is a continuous function of ρ, we may choose ρ so that

$$\frac{\rho'(\rho-r)}{r(\rho'-r)}[2T(\rho',f)+C] = 1.$$

For this choice of ρ, we thus have

$$N(\rho) - N(r) \leq 1 \quad \text{and}$$
(18.13)
$$(\rho - r) = \frac{r(\rho'-r)}{\rho'[2T(\rho',f)+C]}$$

and

$$(\rho' - \rho) = (\rho' - r)\left[1 - \frac{r}{\rho'[2T(\rho',f)+C]}\right].$$

Hence,

$$\log^+ \frac{1}{\rho-r} \le \log^+ \rho' + \log^+ T(\rho',f) + \log^+ \frac{1}{\rho'-r}.$$

Therefore,

(18.14) $$\log^+ \frac{1}{\rho-r} \le \log^+ \rho' + \log^+ T(\rho',f) + \log^+ \frac{1}{\rho'-\rho}.$$

Also, we see that

$$\log^+ \frac{1}{\rho'-\rho} \le \log^+ \frac{1}{\rho'-r} + \log^+ \frac{1}{1 - \frac{r}{\rho'[2T(\rho',f)+C]}}$$

$$\le \log^+ \frac{1}{\rho'-r} + \log^+ \frac{1}{1 - \frac{1}{2T(\rho',f)+C}}.$$

Hence,

(18.15) $$\log^+ \frac{1}{\rho'-\rho} \le \log^+ \frac{1}{\rho'-r}.$$

Using (18.13), (18.14), and (18.15) in (18.12), we have

(18.16) $$m\left(r, \frac{f'}{f}\right) \le 4\log^+ \rho' + 4\log^+ T(\rho',f) + 3\log^+ \frac{1}{\rho'-r},$$

which proves the lemma.

We will use the notation $\|f(x) \le g(x)$ to mean $f(x) \le g(x)$ with Borel exceptions (i.e., a set of finite length).

Proposition.

$$\|m(r, \frac{f'}{f}) \le 4\log^+ r + 8\log^+ T(r,f).$$

Proof. Taking $\rho = r + \frac{1}{\log^+ T(r,f)}$ in the lemma, we have

$$m\left(r, \frac{f'}{f}\right) \le 4\log^+ \left(r + \frac{1}{\log^+ T(r,f)}\right) + 3\log^+ \log^+ T(r,f)$$

$$+ 4\log^+ T\left(r + \frac{1}{\log^+ T(r,f)}, f\right).$$

Hence, by the Borel Lemma, we get

$$\|m\left(r, \frac{f'}{f}\right) \le 4\log^+ r + 8\log^+ T(r,f).$$

Corrollary. $\|S(r) = O(\log^+ T(r,f)) + O(\log^+ r).$

Proof. This follows immediately from the proposition since $S(r)$ is a finite sum of terms of the form $m(r, \frac{f'}{f})$.

19

"Two Constant" Theorems and the Phragmén-Lindelöf Theorems

The Two Constant Theorem. *Suppose that f is holomorphic in $\mathbb{D} = \{z : |z| < 1\}$ and continuous in $\bar{\mathbb{D}}\backslash\{1\}$. Suppose further that $|f| \leq N$ in $\mathbb{D}$ and $|f| \leq M$ in $\partial\mathbb{D}\backslash\{1\}$. Then $|f| \leq M$ in $\bar{\mathbb{D}}\backslash\{1\}$.*

First Proof. Choose $\alpha > 0$ and let

$$g(z) = \left(\frac{1-z}{2}\right)^{\alpha} f(z)$$

for a suitable branch of $\left(\frac{1-z}{2}\right)^{\alpha}$. Now g is analytic in $\mathbb{D}$ and continuous in $\bar{\mathbb{D}}$ if we define $g(1) = 0$. By the maximum modulus theorem,

$$\sup_{z \in \mathbb{D}}|g(z)| = \max_{z \in \partial\mathbb{D}}|g(z)|.$$

But $|g(z)| \leq M$ for $z \in \partial\mathbb{D}$. Hence, $|g(z)| \leq M$ for $z \in \mathbb{D}$, and so

$$|f(z)| \leq M \left|\frac{2}{1-z}\right|^{\alpha} \quad \text{for} \quad z \in \mathbb{D}.$$

Now let $\alpha \to 0$ to get the result.

Second Proof. This proof uses the Cauchy integral formula. Since there exists a linear fractional transformation $U(z) = e^{i\theta}\frac{z-z_0}{1-\bar{z}_0 z}$ taking an arbitrary point $z_0 \in D$ into 0, and mapping $\bar{\mathbb{D}}/\{1\}$ conformally onto $\bar{\mathbb{D}}/\{1\}$, we need only prove that $|f(0)| \leq M$. But choosing θ_0 with $0 \leq \theta_0 < \pi$,

$$|f(0)| = \left|\frac{1}{2\pi}\int_{|w|=R}\frac{f(w)}{w}dw\right| \leq \frac{1}{2\pi}\int_{(1)}|f(re^{i\theta})|d\theta + \frac{1}{2\pi}\int_{(2)}|f(re^{i\theta})|d\theta,$$

where $R < 1$ and

$$\int_{(1)} = \int_{\theta_0 \le |\theta| < \pi}, \qquad \int_{(2)} = \int_{|\theta| \le \theta_0}.$$

But

$$\frac{1}{2\pi} \int_{(2)} \le \frac{\theta_0}{\pi} N,$$

and as $R \to 1^-$

$$\frac{1}{2\pi} \int_{(1)} \le (M + 0(1)) \left(\frac{2\pi - 2\theta_0}{2\pi} \right).$$

Hence, letting $R \to 1^-$,

$$|f(0)| \le \frac{\theta_0}{\pi} N + M \left(1 - \frac{\theta_0}{\pi} \right).$$

Now letting $\theta_0 \to 0$ we again obtain the desired result.

Third Proof. By Jensen's Theorem,

$$\log |f(0)| \le \frac{1}{2\pi} \int_{-\pi}^{\pi} \log |f(re^{i\theta})| d\theta.$$

The argument of the second proof works here, too.

Fourth Proof. By a conformal mapping, we may replace the unit disk by the upper half-plane. We have $|f| \le M$ on the real axis and $|f| \le N$ above the real axis, and must conclude that $|f| \le M$ above the real axis. Choose $A > 0$ and let

$$g(z) = \left(\frac{A}{z + iA} \right) f(z).$$

Note that

$$\left| \frac{A}{z + iA} \right| \le \frac{A}{|z|}$$

and that

$$|g(x)| \le M \quad \text{for all real } x.$$

Hence, if R is large enough, we see by the maximum modulus theorem that $|g(z)| \le M$ for z in the semicircular region bounded by the real axis and the arc $|z| = R$, $0 \le \arg z \le \pi$. Hence, $|g(z)| \le M$ in the upper half-plane. Letting $A \to \infty$, we complete the fourth proof.

A Phragmén-Lindelöf Theorem. *Let $f(z)$ be holomorphic inside an angular region of opening π/α, where $\alpha > 1$, and continuous on the closure of the region. Suppose that $|f(z)| \le M$ on the boundary and that*

$$|f(z)| \le K e^{|z|^\beta}$$

in the region, for some constant β with $\beta < \alpha$. Then $|f(z)| \le M$ in the region.

Proof. Without loss of generality, we may suppose that the region is given by $|\arg z| < \frac{\pi}{2\alpha}$. Choose $\epsilon > 0$, and choose γ with $\beta < \gamma < \alpha$. Let

$$F(z) = e^{-\epsilon z^\gamma} f(z),$$

so that

$$|F(re^{i\theta})| = \exp\{-\epsilon r^\gamma \cos \gamma\theta\}|f(z)|.$$

For $z = re^{i\theta}$ in the closed angle, we have $\exp\{-\epsilon r^\gamma \cos \gamma\theta\} < 1$ so that $|F(z)| \le |f(z)|$ there. Notice also that by using the estimate on f inside the region we have

$$|F(re^{i\theta})| \le \exp\{-\epsilon R^\gamma \cos \gamma\theta\} K \exp\{R^\beta\},$$

so that for R large enough we have

$$|F(Re^{i\theta})| \le M.$$

It follows that

$$|f(z)| \le M \exp\{\epsilon R^\gamma\},$$

and the result follows on letting $\epsilon \to 0$.

Taking for simplicity $\alpha = 1$, the Phragmén-Lindelöf Theorem says that if a function is holomorphic and of order less than one in a half-plane and bounded on the boundary, then it is bounded inside by the same bound. The example $f(z) = e^z$ in the right half-plane shows that the order 1 is critical. However, the next result shows that the analogous result holds if we replace the condition "order less than 1" by "growth at most order 1, zero exponential type."

Theorem. *If the hypothesis is changed to read that for each $\delta > 0$*

$$|f(z)| \le K_\delta \exp\{\delta|z|^\alpha\}$$

in the region, then the conclusion follows as before.

Proof. Now let

$$F(z) = \exp\{-\epsilon z^\alpha\} f(z),$$

and the same proof works.

Remark. Analogous results hold for functions holomorphic in other regions and satisfying appropriate growth restrictions. One useful case is the parallel strip. Calderon has used this case in developing a theory of interpolation of Banach spaces that can be applied in the fields of harmonic analysis and partial differential equations.

20
The Pólya Representation Theorem

The Pólya Representation Theorem plays a central role in the theory of entire functions of exponential-type. We give a somewhat augmented version of this theorem.

Before proceeding with the theorem, it will be necessary to discuss convex sets. We say that a set E (in the complex plane) is convex if E contains the line segment joining any two points. That is, if $z_1, z_2 \in E$, then $tz_1 + (1-t)z_2 \in E$ for all $t \in [0,1]$. The intersection of convex sets is again convex.

Definition. Given a set A, the intersection of all half-planes that contain A is called the *closed convex hull* of A and is denoted by $K(A)$.

Definition. A point is an *extreme point* of a set if it is not the midpoint of any line segment contained in the set.

Theorem 20.1. *A compact convex set is the convex hull of the set of its extreme points.*

We omit the proof.

Definition. For any set E, $k(\theta) = \sup\{Re(ze^{-i\theta}) : z \in E\}$ is called the *support function* of E.

We note that $k(\theta)$ measures the directed distance from the origin to the most remote point of the projection of E on the ray $\arg z = \theta$. Note also that if E is empty, then $k = -\infty$. It is easy to show that, for a given set E,

$$K(E) = \{z : Re(ze^{-i\theta}) \le k(\theta) \quad \text{for all} \quad \theta\}.$$

Remark 20.2. If $z_0 = x_0 + iy_0 = r_o e^{i\theta_0}$ and $E = \{z_0\}$, then $k(\theta) = r_0 \cos(\theta - \theta_0) = x_0 \cos\theta + y_0 \sin\theta$.

Remark 20.3. Let E be a circle with center at 0 and radius R. Then $k(\theta) = R$ for all θ.

Remark 20.4. Let E be the line segment $[x_0, x_1]$, $x_0 < x_1$. Then $k(\theta) = x_1 \cos \theta$ if $-\frac{\pi}{2} \leq \theta \leq \frac{\pi}{2}$ and $k(\theta) = x_0 \cos \theta$ if $\frac{\pi}{2} \leq \theta \leq \frac{3\pi}{2}$. In particular, if $x_0 = -x_1$ and $x_1 = R$, then $k(\theta) = R|\cos \theta|$. If E is the vertical line segment $[-iR, iR]$, then $k(\theta) = R|\sin \theta|$.

Remark 20.5. If E_1 has support function k_1 and E_2 has support function k_2, then $E_1 + E_2 = \{z_1 + z_2 : z_1 \in E_1, z_2 \in E_2\}$ has support function $k_1 + k_2$. From Remark 20.3, it follows then that the rectangle with vertices $(\pm R_1, \pm i R_2)$ has support function $k(\theta) = R_1|\cos \theta| + r_2|\sin \theta|$.

Remark 20.6. The convex hull of $E_1 \cup E_2$ has support function $k = \max\{k_1, k_2\}$.

Remark 20.7. If E_1 with support function k_1 is translated, so that the point originally at 0 is moved to $z_0 = x_0 + iy_0$, then the support function of the translated set is $k(\theta) + x_0 \cos \theta + y_0 \sin \theta$.

Definition. Let $\mathbf{H}_0(\infty)$ be the class of all functions that are holomorphic near ∞ and that vanish at ∞.

Let f be an entire function of exponential-type and write $f(z) = \sum \left(\frac{a_n}{n!}\right) z^n$. The *Borel transform* Φ of f, defined by $\Phi(w) = \sum a_n \frac{1}{w^{n+1}}$, belongs to H_0. Each function in H_0 is the Borel transform of a unique f. We have seen (Corollary to Proposition 11.5) that the type of f is the radius of convergence of the series $\sum a_n \frac{1}{w^{n+1}}$.

In Proposition 11.7, we saw that $\Phi(w) = \int_0^\infty f(t) e^{-tw} dt$, in the sense of analytic continuation. We also saw that

$$f(z) = \frac{1}{2\pi i} \int_\Gamma \Phi(w) e^{zw} dw,$$

where Γ is a rectifiable curve that winds once around the singularities of Φ. We call this the Pólya integral representation formula.

Definition. Let $S(\Phi)$ be the set of singular points of Φ, and let k be its supporting function. We call $S(\Phi)$ the *conjugate indicator diagram* of f. Sometimes this name is used for $S^*(\Phi)$, the closed convex hull of $S(\Phi)$. We will write $D(f) = S^*(\Phi)$.

Definition. The *indicator function of f* is

$$h(\theta) = h_f(\theta) = \limsup_{r \to \infty} \; \frac{1}{r} \log |f(re^{i\theta})|.$$

We now state an important part of the Pólya Representation Theorem.

Theorem 20.8. $h(\theta) = k(-\theta)$ *for all* θ.

The proof of Theorem 20.17 is contained in an appendix at the end of this chapter. We now make some remarks to illustrate this theorem.

Remark 20.9. If f is of zero-type, then $h = 0$ so that $k = 0$, and hence 0 is the only possible singularity of Φ.

Remark 20.10. Denoting by $\tau(f)$ the type of f, we have $\tau(f) = \max h(\theta)$.

Remark 20.11. If $f(z) = e^{az}$, where a is real, then

$$h(\theta) = \limsup_{r\to\infty} \frac{1}{r} \log |e^{az}| = \limsup_{r\to\infty} \frac{1}{r} ar \cos \theta = a \cos \theta.$$

On the other hand, $\Phi(w) = \sum a^n \frac{1}{w^{n+1}} = (w - a)^{-1}$ and so $S(\Phi) = \{a\}$. Thus, $k(\theta) = a \cos \theta$ and we do have $h(\theta) = k(-\theta)$.

Remark 20.12. If $f(z) = e^{iz}$, we have $h(\theta) = -\sin \theta$ and $\Phi(w) = (w-i)^{-1}$. Hence, $k(\theta) = \sin \theta$ since $S(\Phi) = \{i\}$. Note again that $h(\theta) = k(-\theta)$.

Remark 20.13. Suppose $f(z) = \sum a_n e^{\lambda_n z}$, a finite sum with distinct λ_n and nonzero a_n. Then $\Phi(w) = \sum a_n (w-\lambda_n)^{-1}$ so that $S(\Phi) = \{\lambda_n\}$. Note that only the extreme points of $\{\lambda_n\}$ affect $h(\theta)$. If the λ_n lie on a straight line, only the endpoints affect $h(\theta)$.

Remark 20.14. It is easy to see that $h_{fg} \leq h_f + h_g$, so that $D(fg) \subseteq D(f) + D(g)$. An interesting problem that has applications in harmonic analysis is that of finding suitable conditions under which $D(fg) = D(f) + D(g)$.

Let M_0 be the class of all Borel measures of compact support.

Definition. If $d\mu \in M_0$, its Laplace transform $d\mu^\wedge$ is defined by

$$d\mu^\wedge(z) = \int e^{-zw} d\mu(w).$$

It is easy to see that $d\mu^\wedge(z)$ is an entire function of exponential-type for

$$\frac{d}{dz} d\mu^\wedge(z) = \int (-w) e^{-zw} d\mu(w)$$

(as can be verified on differentiating "by hand") so that $d\mu^\wedge$ is entire. And

$$|d\mu^\wedge(z)| \leq \int |e^{zw}| |d\mu(w)| \leq e^{R|z|} \int |d\mu(w)|,$$

where R is chosen so that the support of $d\mu$ lies in the disk of radius R centered at 0.

Definition. We write $d\mu \sim d\nu$ to mean that $d\mu^\wedge = d\nu^\wedge$.

It is not hard to show that $d\mu \sim d\nu$ if and only if $\int f d\mu = \int f d\nu$ for each entire function f, or for each entire function f of exponential-type. It is clear that $\sim$ is an equivalence relation.

If we take $d\mu = dz|_\Gamma$, where Γ is a circle, then $d\mu^\wedge(z) = \int_\Gamma e^{-zw} dw = 0$ by the Cauchy Theorem. Hence, $d\mu \sim 0$ even though $d\mu \neq 0$. We shall see that for any $d\mu \in m_0$, $d\mu \sim d\nu$, where $d\nu = \Phi(-w)(-dw)|_\Gamma$, where Φ is the Borel transform of $d\mu^\wedge$ and Γ is any curve that winds once around $S(\Phi)$.

Definition. $[d\mu]$ is the class of measures equivalent to $d\mu$.

Definition. M_0' is the class of all $[d\mu]$ for $d\mu \in M_0$.

Definition. If f is continuous in the plane and $d\mu$ in M_0, we define the *convolution* $f * d\mu$ by

$$(f * d\mu)(z) = \int f(w - z) d\mu(w).$$

Definition. If $d\mu$ and $d\nu$ are in M_0, we define the convolution $d\mu * d\nu$ by

$$f * (d\mu * d\nu) = (f * d\mu) * d\nu.$$

By means of the Riesz Representation Theorem, it is easy to see that the above definition defines $d\mu * d\nu$ as a unique measure in M_0. Indeed, M_0 is an algebra over the complex numbers.

Proposition. *If $d\mu_1 \sim d\mu_2$, then $(d\mu_1 * d\nu) \sim (d\mu_2 * d\nu)$. It thus makes sense to define $[d\mu] * [d\nu] = [d\mu * d\nu]$. M_0' is an algebra over the complex numbers.*

Definition. Let E_0 be the algebra of all entire functions of exponential-type.

Definition. Let E be the space of all entire functions in the topology of uniform convergence on compact sets.

Definition. E^*, the dual of E, is the space of all continuous linear functionals on E.

Now E is a locally convex topological linear vector space. It will appear that each of the spaces M_0', E_0, $H_0(\infty)$ "is" the dual space E^*.

Definition. For $f \in E$ and $[d\mu] \in M_0'$, define the inner product $\langle F_1, [d\mu] \rangle$ by

$$\langle F_1, [d\mu] \rangle = (F * d\mu)(0) = \int F(-z) d\mu(z).$$

Definition. For $F \in E$ and $f \in E_0$, define the inner product $\langle F, f \rangle$ by

$$\langle F, f \rangle = (f(D)F)(0),$$

where $D = \frac{d}{dz}$. This means that if $f(z) = \sum \frac{a_n}{n!} z^n$, then $\langle F, f \rangle = \sum \frac{a_n}{n!} F^{(n)}(0)$.

It is not hard to show that, for each $f \in E_0$ and $F \in E$, the series defining $\langle F, f \rangle$ converges. Indeed, the linear functional λ defined by $\lambda(F) = \langle F, f \rangle$ is a continuous linear functional on E. The same is true for $\lambda(F) = \langle F, d\mu \rangle$.

Definition. For $F \in E$ and $\Phi \in H_o(\infty)$, define the inner product $\langle F, \Phi \rangle$ by

$$\langle F, \Phi \rangle = \frac{1}{2\pi i} \int_\Gamma \Phi(w) F(w) dw,$$

where Γ is any curve that winds once around $S(\Phi)$

Again, $\lambda(F) = \langle F, \Phi \rangle$ defines a continuous linear functional on E.

Theorem 20.15. *Let $[d\mu] \in M_0'$, $f \in E_0$, $\Phi \in H_o(\infty)$ be related by $f = d\mu^\wedge$ and Φ be the Borel transform of f. Then $d\mu, f$, and Φ give rise to the same functional in E^*. And given any functional in E^*, there is a unique $[d\mu]$ (or f or Φ) that gives rise to it.*

To say that f gives rise to λ is to say that $\lambda(F) = \langle F, f \rangle$ for each $F \in E$, with a similar definition for $d\mu$ or Φ. We omit the proof of the theorem since much of it is straightforward.

Remark. To illustrate the theorem, take $f = e^z$ so that $d\mu$ is the unit point mass at 1 and $\Phi(w) = (w-1)^{-1}$. Then $f(D) = e^D$, so that by Taylor's theorem

$$(f(D)F)(0) = F(0) + F'(0) + \frac{1}{2!} F''(0) + \cdots = F(1).$$

Note that $\int F(-z) d\mu(z) = F(1)$ also. And, by Cauchy's Theorem,

$$\frac{1}{2\pi i} \int_\Gamma F(w) \Phi(w) dw = \frac{1}{2\pi i} \int \frac{F(w)}{w-1} dw = F(1).$$

Summarizing the results of this section so far, we have the following omnibus theorem. We call it the Pólya Representation Theorem, although the name is not entirely appropriate.

The Pólya Representation Theorem. *Let E_0 be the algebra of all entire functions of exponential-type and M_0' the convolution algebra of equivalence classes of Borel measures of compact support, where $d\mu \sim d\nu$ means that $\int f d\mu = \int f d\nu$ for each entire function f. The Laplace transform is defined by $d\mu^\wedge(z) = \int e^{-zw} d\mu(w)$; $d\mu \sim d\nu$ if and only if $d\mu^\wedge = d\nu^\wedge$, so* *it makes sense to talk of $[d\mu]^\wedge = d\mu^\wedge$, where $[d\mu]$ is the equivalence class that contains $d\mu$. The Laplace transform is an isomorphism of M_0' onto E_0. To invert the Laplace transform, i.e., given $f \in E_0$, to find $d\mu \in M_0$ such that $d\mu^\wedge = f$, take $d\mu(z) = \Phi(-z)(-dz)|_\Gamma$, where Φ is the Borel transform of f and Γ winds once around the set $S(\Phi)$ of singularities of Φ. The indicator diagram of f, $D(f)$, is defined as the convex hull of $S(\Phi)$. If $h(\theta) = \limsup\limits_{r \to \infty} \frac{\log |f(re^{i\theta})|}{r}$ is the indicator function of f, then $h(\theta) = k(-\theta)$, where k is the support function of $D(f)$. If E is the space of all entire functions in the topology of uniform convergence on compact sets, then both E_0 and M_0' are the dual space of E, where, if $f = d\mu^\wedge$, then for $F \in E$*

$$\langle F, f \rangle = (f(D)F)(0) = \langle F, d\mu \rangle = \int F(-z) d\mu(z).$$

We now turn to some applications of this theorem.

[Note. We repeatedly use conventional, but strictly incorrect, phrases like "Γ winds once around the set $S(\Phi)$ of singularities of Φ." A more correct wording is "Γ has winding number -1 around ∞, and lies in a connected and simply connected open set containing ∞ in which Φ is analytic."]

Definition. For an entire function f and a complex number λ, let f_λ be the entire function defined by

$$f_\lambda(z) = f(z + \lambda) \quad \text{for all } z.$$

Lemma 20.16. *For any entire function f, $\tau(f_\lambda) = \tau(f)$.*

Proof. With $M(r : f) = \max\{|f(z)| : |z| = r\}$, we have

$$M(r : f_\lambda) \leq M(r + |\lambda| : f)$$

so that

$$\frac{1}{r} \log M(r : f_\lambda) \leq \frac{1}{r} M(r + |\lambda| : f) = (1 + 0(1)) \frac{1}{r + |\lambda|} \log M(r + |\lambda| : f).$$

Hence,

$$\tau(f_\lambda) \leq \tau(f).$$

Similarly,

$$\tau(f) \leq \tau(f_\lambda) \quad \text{since} \quad f(f_\lambda)_{-\lambda}.$$

Lemma 20.17. *If f and g are entire functions of exponential-type, then $D(f) = D(g)$ if and only if*

$$\tau(f(z)e^{az}) = \tau(g(z)e^{az})$$

for each complex number a.

Proof. Clearly, if $D(f) = D(g)$, then $\tau(f(z)e^{az}) = \tau(g(z)e^{az})$. To prove the converse, it is enough to compute, say, $h(0)$ from $\tau(f(z)e^{az})$. We show that

$$h(0) = \lim_{a \to +\infty} \tau(f(z)e^{az}) - a.$$

Let D_a be the indicator diagram of $f(z)e^{az}$. Then $D_a = a + D(f)$, that is, D_a is $D(f)$ translated by a. Choose $B \leq \max\left(|h_f\left(\frac{\pi}{2}\right)|, |h_f\left(-\frac{\pi}{2}\right)|\right)$.

Now when a is large, D_a lies to the right of the origin, and D_a lies in the strip $\{z = x + iy : |y| \leq B; x \leq a + h_f(0)\}$. D_a also contains the point $(a + h(0), 0)$. Hence, if $\tau_a = \tau(f(z)e^{az})$, then

$$\tau_a \geq a + h(0)$$

and

$$\tau \leq \{(a + h(0))^2 + B^2\}^{\frac{1}{2}},$$

so that

$$\tau_a - (a + h(0)) = o(1).$$

□

Lemma 20.18. *If f is an entire function of exponential-type, then for each complex number λ,*

$$D(f) = D(f_\lambda).$$

Proof. We use Lemma 20.17, with $g = f_\lambda$. Note that

$$e^{az}g(z) = e^{az}f(z + \lambda) = e^{-a\lambda}(e^{a(z+\lambda)}f(z + \lambda)).$$

So if we let $F(z) = e^{az}f(z)$, then $e^{az}g(z) = CF_\lambda(z)$, where C is a nonzero constant. Since $\tau(F_\lambda) = \tau(f)$, we have $\tau(e^{az}f(z)) = \tau(e^{az}f_\lambda(z))$, and the result follows.

The Pólya Theorem has the following corollary.

Proposition. *If $h\left(\frac{\pi}{2}\right) < 0$ and $h\left(-\frac{\pi}{2}\right) < 0$, then $f = 0$.*

Proof. $h\left(\frac{\pi}{2}\right) < 0$ implies that $D(f)$ lies above the real axis and $h\left(-\frac{\pi}{2}\right) < 0$ implies that $D(f)$ lies below the real axis. Hence, $D(f)$ is empty and, consequently, Φ has no singularities.

Since $\Phi(\infty) = 0$, it follows from Liouville's Theorem that $\Phi = 0$, and hence $f = 0$.

The Pólya Representation Theorem provides an alternate proof of Carlson's Theorem presented in Chapter 12.

Theorem (F. Carlson, 1914) **20.19.**   *Suppose f is an entire function of exponential-type with $\tau(f) < \pi$, and suppose that $f(n) = 0$, $n = 0, \pm 1, \pm 2, \ldots$. Then $f = 0$.*

Proof. Consider the function $g(z) = f(z)/\sin \pi z$. It is clear that g is entire. Since

$$T(r, F/G) \leq T(r, F) + T(r, G),$$

g is of exponential-type. Now $h_g\left(-\frac{\pi}{2}\right) < 0$, and $h_g\left(\frac{\pi}{2}\right) < 0$, so that by the above proposition, $g = 0$. Hence $f = 0$.

Remark 20.20. It is clear from the proof that the hypothesis that $\tau(f) < \pi$ can be relaxed to $h_f(\pm \pi/2) < \pi$. The condition $\tau(f) < \pi$ is sharp, since $\sin \pi z$ is of type π. The Carlson Theorem says, roughly, that an entire function of exponential-type must grow fast in a direction at right angles to its zeros. Later in this section, we shall get a stronger version of Carlson's Theorem. Chapter 22 is devoted to proving an extremely strong generalization of it.

Definition. Let k denote the set of all entire functions f of exponential-type such that $h_f(\pm \pi/2) < \pi$.

Definition. A sequence $\{A_n\}$, $n = 0, 1, 2, \ldots$ is said to be k-admissible provided that there exists an $f \in k$ such that $f(n) = A_n$, $n = 0, 1, 2, \ldots$.

We now give a characterization, due to Buck, of k-admissible sequences.

Theorem (Horseshoe Theorem) **20.21.**   *The sequence $\{A_n\}$ is called k-admissible if and only if $F(z) = \sum A_n z^n$ is holomorphic at 0 and ∞ and in some neighborhood of the negative real axis as well.*

Proof. Suppose first that $\{A_n\}$ is k-admissible, so that $A_n = f(n)$ for some $f \in k$. Then, if z is small,

$$F(z) = \sum f(n) z^n = \sum \left(\frac{1}{2\pi i} \int_\Gamma \Phi(w) e^{nw} \, dw\right) z^n = \frac{1}{2\pi i} \int_\Gamma \frac{\Phi(w)}{1 - ze^w} \, dw.$$

We choose Γ to be a rectangular path that winds around $S(\Phi)$ such that Γ lies in the strip

$$S_\pi = \{z = x + iy : |y| < \pi\}.$$

It follows that $F(z)$ can be analytically continued in the complement of the image of $S(\Phi)$ under the mapping e^{-w}, and that $F(\infty) = 0$.

To prove the other half of the theorem, suppose that F satisfies the requirements of the theorem. We may write

$$A_n = \frac{1}{2\pi i} \int_\Lambda \frac{F(z)}{z^n} \frac{dz}{z},$$

where Λ is a circle around O. But Λ is homotopic, in the complement of $S(F)$, the set of singularities of F, to the path Λ_R illustrated in the figure below, where Λ'_R is a circle of radius R.

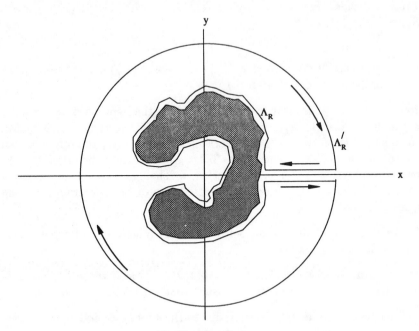

The Horseshoe Contour

Hence,

$$A_n = \frac{1}{2\pi i} \int_{\Lambda_R} \frac{F(z)}{z^n} \frac{dz}{z}.$$

Now

$$\int_{\Lambda'_R} \to 0 \quad \text{as} \quad R \to \infty, \quad \text{since} \quad F(\infty) = 0.$$

Hence,

$$A_n = \frac{1}{2\pi i} \int_{\Lambda^*} \frac{F(z)}{z^n} \frac{dz}{z},$$

where Λ^* is any curve that winds around $S(F)$ with multiplicity -1. We thus are led to define

$$f(w) = \frac{1}{2\pi i} \int_{\Lambda^*} \frac{F(z)}{z^w} \frac{dz}{z}$$

for an appropriate branch of z^w on Λ^*. It is easy to check that $f \in k$, and the proof is complete.

We now may prove the following extension of Carlson's Theorem.

Theorem 20.22. *Suppose that $f \in k$ and that $f(n) = 0$ for $n = 0, 1, 2, \ldots$. Then $f = 0$.*

Proof. We have

$$F(z) = \sum f(n) z^n = \frac{1}{2\pi i} \int_\Gamma (1 - ze^w)^{-1} \Phi(w) dw.$$

By hypothesis, $F = 0$. Hence,

$$zF(z) = \frac{1}{2\pi i} \int_\Gamma \frac{z}{1 - ze^w} \Phi(w) dw = 0.$$

On letting $z \to \infty$, we have

$$-\frac{1}{2\pi i} \int e^{-w} \Phi(w) dw = 0.$$

Since $f(z) = \frac{1}{2\pi i} \int_\Gamma e^{zw} \Phi(w) dw$, we have $f(-1) = 0$.

Now we use the fact that $f \in k$; thus $f_{-1} \in k$, where $f_{-1}(z) = f(z - 1)$. We may apply the same argument to f_{-1} to conclude that $f_{-1}(-1) = 0$, that is, $f(-2) = 0$. Repeating the argument, we get $f(n) = 0$ for $n = 0, \pm 1, \pm 2, \ldots$, so that $f = 0$ by Carlson's Theorem.

We may prove several theorems about k-admissible sequences by means of the above characterization using classical theorems of function theory that relate the properties of a function to its power series coefficients.

Theorem 20.23. *If $\{A_n\}$ is k-admissible and $|A_n|^{1/n} \to 0$, then $A_n = 0$ for each n.*

The proof is trivial.

Theorem (Pringsheim). *Let $f(z) = \sum \alpha_n z^n$ have nonnegative coefficients and radius of convergence R. Then $z = R$ is a singular point of f.*

Theorem 20.24. *If $\{A_n\}$ is admissible and $(-1)^n A_n \geq 0$ for $n = 0, 1, 2, \ldots$, then $A_n = 0$ for $n = 0, 1, 2, \ldots$.*

Proof. $F(-z) = \sum (-1)^n A_n z^n$ has nonnegative coefficients, but no positive real number is a singularity of F.

Hadamard Gap Theorem. *If $f(z) = \sum \alpha_n z^n$ with $\alpha_n = 0$ except for $n = n_k$, where $\varliminf \frac{n_{k+1}}{n_k} > 1$, then every point of the circle of convergence of f is a singular point of f.*

Theorem 20.25. *If $\{A_n\}$ is k-admissible and $A_n = 0$ except for $n = n_k$ with $\varliminf \frac{n_{k+1}}{n_k} > 1$, then $A_n = 0$ for $n = 0, 1, 2, \ldots$.*

This is a simple consequence of the Hadamard Gap Theorem.

Fabry Gap Theorem. *If $f(z) = \sum \alpha_n z^n$ with $\alpha_n = 0$ except for $n = n_k$ and $k^{-1} n_k \to \infty$, then every point on the circle of convergence of f is a singular point of f.*

Theorem (Szegö). *Suppose that $f(z) = \sum \alpha_n z^n$, where the α_n lie in some finite set. Either $|z| = 1$ is a natural boundary of f or f is a rational function, and the α_n are eventually periodic.*

As a corollary we have

Theorem 20.26. *If $\{A_n\}$ is k-admissible and the A_n lie in some finite set, then the A_n are eventually periodic.*

Appendix

The proof that $h(\theta) = k(-\theta)$ is presented in this section. The actual proof of this assertion is fairly simple, but we prefer to give some of the background concerning supporting functions of convex sets. First, we give a simple necessary and sufficient condition that a function $h(\theta)$ should be the supporting function of a nonempty compact convex set. The condition is that the function should be "subsinusoidal." Next, we prove that if $h(\theta)$ is the indicator function of an entire function of exponential type, then $h(\theta)$ is subsinusoidal. Finally, we show that $h(\theta)$ is the supporting function of the conjugate of the conjugate indicator diagram. From now on, when we speak of a "function of θ," we mean a function that is 2π-periodic; and when we speak of a "supporting function," we mean a supporting function of a nonempty compact convex set.

Our treatment is a combination of the treatments in Pólya [31] and Boas [5].

Definition. A function $H(\theta)$ is a *sinusoid* if it has the form $H(\theta) = a\cos\theta + b\sin\theta$.

Remark. Given $\theta_1 \neq \theta_2$ and real numbers h_1 and h_2, there is a unique sinusoid H such that $H(\theta_1) = h_1$ and $H(\theta_2) = h_2$. We call H the *interpolating sinusoid:* It is given by ($0 < \theta_2 - \theta_1 < \pi$)

$$(20.1) \qquad H(\theta) = h_1 \frac{\sin(\theta_2 - \theta)}{\sin(\theta_2 - \theta_1)} + h_2 \frac{\sin(\theta - \theta_1)}{\sin(\theta_2 - \theta_1)}.$$

Definition. Given a function $h(\theta)$ and $\theta_1 \neq \theta_2$, we call H the *interpolating sinusoid of h* if H is given by (20.1) with $h_1 = h(\theta_1)$ and $h_2 = h(\theta_2)$.

Definition. A function $h(\theta)$ is *subsinusoidal* if it is majorized by each of its interpolating sinusoids, that is,

$$(20.2) \qquad h(\theta_2) \leq h(\theta_1)\frac{\sin(\theta_3 - \theta_2)}{\sin(\theta_3 - \theta_1)} + h(\theta_3)\frac{\sin(\theta_2 - \theta_1)}{\sin(\theta_3 - \theta_1)}$$

whenever $\theta_1 \leq \theta_2 \leq \theta_3$ with $0 < \theta_3 - \theta_1 < \pi$.

Remark. The theory of subsinusoidal functions has some similarity to the theory of convex functions.

Remark. If h is subsinusoidal, and if H is sinusoidal and $H(\theta_1) \geq h(\theta_1)$, $H(\theta_2) \geq h(\theta_2)$, then $H(\theta) \geq h(\theta)$ if $\theta_1 \leq \theta \leq \theta_2$ with $0 < \theta_2 - \theta_1 < \pi$.

Remark. The sum of two subsinusoidal functions is subsinusoidal.

Remark. That h is subsinusoidal is equivalent to the assertion that the point $h(\theta)e^{i\theta}$ does not lie outside the circle that passes through the points 0, $h(\theta_1)e^{i\theta_1}$, and $h(\theta_2)e^{i\theta_2}$, where θ_1, θ_2, and θ are in the specified range. This geometric interpretation can be used to supply geometric proofs of some of the subsequent results.

Problem. Suppose that h_1 is subsinusoidal, h_2 is supersinusoidal, and $h_1(\theta) \geq h_2(\theta)$ for all θ. Does there exist a sinusoidal function H such that $h_1(\theta) \geq H(\theta) \geq h_2(\theta)$ for all θ?

Lemma 20.27. *Suppose that h is subsinusoidal, that $\theta_1 < \theta_2 < \theta_3$, that $0 < \theta_3 - \theta < \pi$, and that $H(\theta)$ is a sinusoid such that $h(\theta_1) \leq H(\theta_1)$ and $h(\theta_2) \geq H(\theta_2)$. Then $h(\theta_3) \geq H(\theta_3)$.*

Proof. Suppose $\delta > 0$ and $h(\theta_3) < H(\theta_3) - \delta$. Let H_δ be the sinusoid such that

$$H_\delta(\theta_1) = H(\theta_1), \quad H_\delta(\theta_3) = H(\theta_3) - \delta.$$

Then $H_\delta(\theta_2) < H(\theta_2)$, since

$$H_\delta(\theta) = H(\theta) - \delta \frac{\sin(\theta - \theta_1)}{\sin(\theta_3 - \theta_1)}.$$

Since h is subsinusoidal and we have

$$h(\theta_1) \leq H_\delta(\theta_1), \quad h(\theta_2) \leq H_\delta(\theta_2),$$

it follows that

$$h(\theta_2) \leq H_\delta(\theta_2) < H(\theta_2),$$

which is impossible.

Lemma 20.28. *A function $h(\theta)$ is subsinusoidal if and only if*

$$(20.3) \quad h(\theta_1)\sin(\theta_3 - \theta_2) + h(\theta_2)\sin(\theta_1 - \theta_3) + h(\theta_3)\sin(\theta_2 - \theta_1) \geq 0$$

whenever $\theta_1 < \theta_2 < \theta_3$; $\theta_2 - \theta_1 < \pi$; $\theta_3 - \theta_2 < \pi$.

Proof. Clearly, (20.3) is equivalent to (20.2) if $\theta_3 - \theta_1 < \pi$. To prove (20.3) in general, choose θ_4 so that $\theta_2 < \theta_4 < \theta_1 + \pi$ and let $H(\theta)$ be the interpolating sinusoidal for h at θ_1, θ_2. By Lemma 20.27, $h(\theta_4) \geq H(\theta_4)$. Repeating this argument with θ_2, θ_4, θ_3 we get $h(\theta_3) \geq H(\theta_3)$. Now

$$h(\theta_1)\sin(\theta_3 - \theta_2) + h(\theta_2)\sin(\theta_1 - \theta_3) + H(\theta_3)\sin(\theta_2 - \theta_1) = 0,$$

but $\sin(\theta_2 - \theta_1) > 0$ and $h(\theta_3) \geq H(\theta_3)$, so the result follows.

Lemma 20.29. *If h is subsinusoidal, then it is continuous and even has left and right derivatives at each point. The left derivative is never greater than the right derivative.*

Proof. Choose θ and suppose without loss of generality that $h(\theta) < 0$. Otherwise, consider $h - H$ where H is a sinusoid such that $H(\theta) > h(\theta)$. Choose $\epsilon > 0$, $\delta > 0$ with $\epsilon + \delta < \pi$. Applying (20.2) in turn to the following triples φ_1, φ_2, φ_3:

  (i) $\theta - \epsilon - \delta$, $\theta - \epsilon$, θ
  (ii) $\theta - \epsilon$, θ, $\theta + \epsilon$,
  (iii) θ, $\theta + \epsilon$, $\theta + \epsilon + \delta$,
  we eventually obtain

(20.4)
$$\frac{h(\theta) - h(\theta - \epsilon - \delta)}{\sin(\epsilon + \delta)} < \frac{h(\theta) - h(\theta - \epsilon)}{\sin \epsilon}$$
$$< \frac{h(\theta + \epsilon) - h(\theta)}{\sin \epsilon} < \frac{h(\theta + \epsilon + \delta) - h(\theta)}{\sin(\epsilon + \delta)}.$$

For example, considering the triple (i), we get

$$h(\theta - \epsilon - \delta)\sin \epsilon - h(\theta - \epsilon)sin(\epsilon + \delta) > -h(\theta)\sin \delta$$

so that

$$[h(\theta - \epsilon - \delta) - h(\theta)]\sin \epsilon - [h(\theta - \epsilon) - h(\theta)]\sin(\epsilon + \delta)$$
$$\geq h(\theta)[\sin(\epsilon + \delta) - \sin \epsilon - \sin \delta]$$
$$= -h(\theta)2\sin\frac{\epsilon + \delta}{2}\left[\cos\frac{\epsilon - \delta}{2} - \cos\frac{\epsilon + \delta}{2}\right] > 0.$$

Hence,

$$[h(\theta - \epsilon - \delta) - h(\theta)]\sin \epsilon > [h(\theta - \epsilon) - h(\theta)]\sin(\epsilon + \delta),$$

which proves the first inequality of (20.4). The others are proved in a similar way. Now (20.4) is precisely the assertion that $\frac{h(\theta + x) - h(\theta)}{\sin x}$ is an increasing function of x when x is small, from which all of the assertions of the lemma follow easily.

Lemma 20.30. *If $h'(\theta)$ denotes the right derivative of $h(\theta)$, then*

$$h(\theta)\cos(\varphi - \theta) + h'(\theta)\sin(\varphi - \theta) - h(\varphi) \leq 0$$

for each pair θ, φ with $\theta - \pi < \varphi < \theta + \pi$.

Proof. Choose ϵ so that $0 < \epsilon < \pi$ and apply (20.2) to the following triples φ_1, φ_2, φ_3: (i) φ, θ, $\theta + \epsilon$ (ii) θ, $\theta + \epsilon$, φ. This use of (20.2) is admissible if either $\theta - \pi \leq \theta < \lambda$ or $\theta + \epsilon < \varphi < \epsilon + \pi$. We get

(20.5) $h(\theta)\sin(\varphi - \theta - \epsilon) + h(\theta + \epsilon)\sin(\theta - \varphi) + h(\varphi)\sin \epsilon \geq 0,$

which may be rewritten as

(20.6)
$$h(\theta)\frac{\sin(\varphi-\theta)-\sin(\varphi-\theta-\epsilon)}{\sin\epsilon}+\frac{h(\theta+\epsilon)-h(\theta)}{\sin\epsilon}\sin(\varphi-\theta)-h(\varphi)\le 0.$$

Now let $\epsilon\to 0$ to get the assertion of the lemma.

Theorem 20.31. *The function $h(\theta)$ is the supporting function of some nonempty compact convex set K if and only if $h(\theta)$ is subsinusoidal.*

Proof. Suppose first that h supports K, so that for each $z=x+iy\in K$ and any θ_1,θ_3 we have

(20.7)
$$x\cos\theta_1+y\sin\theta_1\le h(\theta_1)$$

(20.8)
$$x\cos\theta_3+y\sin\theta_3\le h(\theta_3).$$

If we now choose $\theta_1,\theta_2,\theta_3$ with $\theta_1<\theta_2<\theta_3$, $\theta_2-\theta_1<\pi$, and $\theta_3-\theta_2<\pi$, we may multiply (20.7) by the positive quantity $\sin(\theta_3-\theta_2)$ and (20.8) by the positive quantity $\sin(\theta_2-\theta_1)$ and then add to get

$$h(\theta_1)\sin(\theta_3-\theta_2)+h(\theta_3)\sin(\theta_2-\theta_1)\ge(x\cos\theta_2+y\sin\theta_2)\sin(\theta_3-\theta_1);$$

but $x\cos\theta+y\sin\theta=h(\theta)$ for some $z=x+iy$ in K, and it follows that h is subsinusoidal.

To prove that if h is subsinusoidal, then it is supporting, let

$$K_\theta=\{z=x+iy:x\cos\theta+y\sin\theta\le h(\theta)\}$$

and then let

$$K=\bigcap_\theta K_\theta.$$

Each K_θ is a closed half-plane in the direction θ. It follows that K is closed and convex. Further, K is bounded since it is contained in the rectangle $K_0\cap K_{\frac{\pi}{2}}\cap K_\pi\cap K_{\frac{3\pi}{2}}$. We now prove that each K_θ contains a point of K on its boundary, thus completing the proof of the theorem. Given θ, then, we must prove that there exists a point $z_\theta=x_\theta+iy_\theta\in K$ for which

(20.9)
$$x_\theta\cos\theta+y_\theta\sin\theta=h(\theta).$$

By the theory of envelopes, we also want

(20.10)
$$x_\theta\sin\theta-y_\theta\cos\theta=-h'(\theta),$$

where we interpret $h'(\theta)$ as the right derivative.

Following this heuristic idea, we let x_θ,y_θ be the simultaneous solution of (20.9) and (20.10). But on applying Lemma 20.30, we get for all φ

$$x_\theta\cos\varphi+y_\theta\sin\varphi\le h(\varphi).$$

It follows that $z_\theta\in K$, and since

$$h(\theta)=\max\{x\cos\theta+y\sin\theta:x+iy\in K\},$$

the result is proved.

Theorem 20.32. *If f is an entire function of exponential type and*

$$h(\theta) = h_f(\theta) = \varlimsup_{r \to \infty} \frac{1}{r} \log |f(re^{i\theta})|,$$

then $h(\theta)$ is subsinusoidal, and $h(\theta)$ is consequently a supporting function.

Proof. The proof is a simple application of the Phragmén-Lindelöf principle. Suppose that $0 < \theta_2 - \theta_1 < \pi$, and let $h_1 = h(\theta_1)$, $h_2 = h(\theta_2)$, and for $\delta > 0$ let H_δ be the interpolating sinusoid for $h_1 + \delta$, $h_2 + \delta$ at θ_1, θ_2. Let $A = A_\delta$ be the complex number for which $H_\delta(\theta) = \Re\{Ae^{-i\theta}\}$, and let $F(z) = f(z)e^{-Az}$ so that if $z = re^{i\theta}$, then

$$|F(z)| = |f(z)| \exp\{-rH_\delta(\theta)\}.$$

Now $F(z)$ is bounded on the rays $z = re^{i\theta_1}$, $z = re^{i\theta_2}$, and of order 1 in the angle between. By the Phragmén-Lindelöf Theorem, $F(z)$ is bounded in the angle $\theta_1 \le \theta \le \theta_2$, and it follows that $h(\theta) \le H(\theta)$ for each θ in this range. Since $\lim_{\delta \to 0} H_\delta(\theta) = H_0(\theta)$, the result is proved.

Theorem 20.33. *If f is an entire function of exponential type and $h(\theta) = h_f(\theta) = \varlimsup_{r \to \infty} \frac{1}{r} \log |f(re^{i\theta})|$, then $h(\theta) = k(-\theta)$ where k is the supporting function for the conjugate indicator diagram.*

Proof. Let $\Phi = \Phi_f$ be the Borel transform of f and let $D = D_f$, the conjugate indicator diagram of f, be the convex hull of the set of singularities of Φ. Let C be a rectifiable curve that winds once around D and that stays within an ϵ-neighborhood of D, where $\epsilon > 0$.

Now

$$f(z) = \frac{1}{2\pi i} \int_C \Phi(w) \exp(zw) \, dw,$$

so that if $z = re^{i\theta}$, then

$$|f(re^{i\theta})| \le A \max_{w \in C} |\exp(zw)|,$$

where A is a constant. Hence,

$$h(\theta) \le \max_{w \in C} \Re(we^{i\theta}).$$

If we now let $\epsilon \to 0$, we see that $h(\theta) \le k(-\theta)$.

In the other direction, it is enough to prove that $h(0) \ge k(0)$ since, if we replace $f(z)$ by $g(z) = f(ze^{i\varphi})$, the general case follows from $h_g(0) \ge k_g(0)$, since $h_g(\theta) = h_f(\theta + \varphi)$, $\Phi_g(w) = e^{i\varphi}\Phi_f(we^{-i\varphi})$ $D_g = e^{i\varphi}D_f$ and $k_g(\theta) = k(\theta - \varphi)$. Now, as we have seen,

$$\Phi(w) = \int_0^\infty f(t)e^{-tw} \, dt \quad \text{for } w > h(0),$$

so that Φ has no singularity to the right of the line $x = h(0)$ and the inequality $h(0) \ge k(0)$ follows.

21
Integer-Valued Entire Functions

An *integer-valued entire function* f is one such that $f(n)$ is an integer for $n = 0, 1, 2, \cdots$. Some examples are

(i) $\sin \pi z$

(ii) 2^z

(iii) any polynomial with integer coefficients.

In this section, we shall mainly follow a paper of Buck [7]. In outline, a certain construction generates a special class R_1 of integer-valued entire functions. We will be concerned with finding growth conditions on an integer-valued entire function f that imply $f \in R_1$. The three examples above belong to R_1.

Definition. We say that an algebraic number α is an algebraic integer if it satisfies a polynomial equation:

$$(21.1) \qquad P(z) = z^n + a_{n-1}z^{n-1} + \cdots + a_0 = 0,$$

where $a_j \in \mathbb{Z}$, $j = 0, 1, \ldots, n-1$. Notice that the coefficient of z^n is 1.

Examples. *Any $n \in \mathbb{Z}$ satisfies $z - n = 0$. And $\pm i$ satisfies $z^2 + 1 = 0$.*

It is not hard to prove that the algebraic integers form a ring. If the integer n in (21.1) is minimal, the other roots are called the *conjugates* of α. The collection of all of the roots of a minimal polynomial is called a *complete set of algebraic conjugates*. It is not hard to show that if α is a root of P (where P is not necessarily minimal), then each conjugate of α is also a root.

Consider now the polynomial

$$Q(x) = 1 + q_1 x + q_2 x^2 + \cdots + q_n x^n,$$

where $q_j \in \mathbb{Z}$ for $j = 1, 2, \ldots, n$. We can write

$$Q(x) = \prod_{j=1}^{n} (1 - \beta_j x),$$

where the β_j run over one or more complete sets of algebraic integers. This may be seen from the fact that the β_j are the roots of the polynomial

$$R(x) = x^n Q\left(\frac{1}{x}\right) = x^n + q_1 x^{n-1} + \cdots + q_n.$$

Now let P be any polynomial with integer coefficients and Q as above. Then

$$\frac{P(x)}{Q(x)} = \sum_{0}^{\infty} b_n x^n, \quad \text{where} \quad b_n \in \mathbb{Z}.$$

This follows since we can write

$$\frac{1}{Q(x)} = \frac{1}{1 + q_1 x + \cdots + q_n x^n} = \frac{1}{1 - Q^*(x)} = 1 + Q^*(x) + [Q^*(x)]^2 + \cdots$$
$$= 1 + B_1 x + B_2 x^2 + \cdots .$$

The B_j are clearly integers. To express the b_j in terms of the β_j, let us first suppose for simplicity that the β_j are distinct, and that $P = 1$. Using partial fractions and writing

$$\frac{1}{1 - \beta_j x} = \sum_{n=1}^{\infty} \beta_j^n x^n,$$

we have

$$\frac{1}{\prod_{1}^{m} (1 - \beta_j x)} = \sum_{j=1}^{m} \frac{E_j}{1 - \beta_j x} = \sum_{j=1}^{m} \sum_{n=0}^{\infty} E_j \beta_j^n x^n,$$

where the E_j are the coefficients in the partial fraction expansion. Hence,

$$b_n = \sum_{j=1}^{n} E_j \beta_j^n,$$

so that it is natural to take

$$f(z) = \sum_{j=1}^{m} E_j \beta_j^z,$$

where β_j^z is suitably defined.

Example

$$Q(z) = 1 + z^2 = (1 + iz)(1 - iz)$$

$$\frac{1}{Q(z)} = \frac{1}{2}\frac{1}{1-iz} + \frac{1}{2}\frac{1}{1+iz} = \frac{1}{2}\{1 + iz + i^2 z^2 + \cdots\}$$

$$+ \frac{1}{2}\{1 + (-i)z + (-i)^2 z^2 + \cdots\}$$

$$b_n = \frac{1}{2}[i^n + (-i)^n]$$

$$\{b_n\} = 1, 0, -1, 0, 1, 0, -1, 0, \ldots$$

$$f(z) = \frac{1}{2}[i^z + (-i)^z] = \cos\frac{\pi}{2}z.$$

In case Q has repeated roots, or if P is not constant, some minor modifications must be made, and in general we have

$$b_n = \sum_{j=1}^{m} P_j(n)\beta_j^n,$$

so that we take

(21.2)
$$f(z) = \sum_{j=1}^{m} P_j(z)\beta_j^z,$$

where the P_j are suitable polynomials (not necessarily integer-valued).

Definition. Let R_1 be the class of functions f constructed above.

Definition. Let R be the class of integer-valued entire functions f of exponential-type for which $h_f(\pm\pi/2) < \pi$.

Our problem is to find additional growth conditions on f that imply $f \in R_1$ if $f \in R$. The conditions will be phrased in terms of the "mapping radius" of certain sets associated with the indicator diagram of f.

Definition. Let S be a simply connected open set containing 0 such that the complement of S contains at least two points. Let φ be the function (whose existence and uniqueness is guaranteed by the Riemann Mapping Theorem) that maps S conformally one-one onto the unit disk $\mathbb{D} = \{z : |z| < 1\}$ and such that $\varphi(0) = 0$ and $\varphi'(0) > 0$. Then φ is the normalized mapping function of S and $\rho(S) = \frac{1}{\varphi'(0)}$ is called the *mapping radius* of S.

We shall need the following deep theorem of Pólya, which we state without proof. The proof may be found in [8].

Theorem. *If $g(z) = \sum_{n=0}^{\infty} b_n z^n$, $b_n \in \mathbb{Z}$, and if g is analytic in a region S containing 0, with $\rho(S) > 1$, then there exist polynomials P and Q, with integer coefficients, $Q(0) = 1$, such that*

$$g = \frac{P}{Q}.$$

We also require a simple lemma on polynomials, whose proof we leave to the reader.

Lemma. *If P and Q are polynomials with integer coefficients and $Q(0) = 1$, then there exist polynomials P_1 and Q_1 with integer coefficients $Q_1(0) = 1$ such that P_1 and Q_1 have no common factors and $P_1/Q_1 = P/Q$.*

Now given $f \in R$, let

$$g(z) = \sum_{n=0}^{\infty} f(n)z^n.$$

As we have seen earlier,

$$g(z) = \frac{1}{2\pi i} \int_{\Gamma} \frac{1}{1 - ze^w} \Phi(w)dw$$

for any curve Γ that winds once around $D(f) = S^*(\Phi)$. Now let S be the complement of the image of $D(f)$ under the mapping e^{-w}, i.e., $S = \mathbb{C} \backslash \exp(-D(f))$.

Now g is analytic in S, so that if $\rho(S) > 1$, then $g = \frac{P}{Q}$, where P and Q are polynomials with integer coefficients and $Q(0) = 1$. By the lemma above, we may suppose that P and Q have no common factors. By the construction that characterizes R_1, we can find a function $f_1 \in R_1$ such that $f_1(n) = f(n)$ for $n = 0, 1, 2, \ldots$. By Carlson's Theorem, if we know that $h_{f_1}\left(\pm\frac{\pi}{2}\right) < \pi$, then we have $f = f_1$, so that $f \in R$. To prove that $h_{f_1}\left(\pm\frac{\pi}{2}\right) < \pi$, we write

$$f_1(z) = \sum_{j=1}^{m} P_j(z)\beta_j^z.$$

By construction, the β_j^{-1} are the roots of Q, so that the β_j^{-1} are the singularities of g, and hence the β_j^{-1} are in the complement of S. Hence, we may write $\beta_j = exp(-\mu_j)$, where $\mu_j \in D(f)$, so that

$$f_1(z) = \sum_{j=1}^{m} P_j(z)\exp(-\mu_j z),$$

and it follows that $h_{f_1}\left(\pm\frac{\pi}{2}\right) < \pi$ since $D(f)$ is interior to the strip $|y| < \pi$. We therefore have proved the next theorem.

Theorem. *If $f \in R$, and if the complement of the image of $D(f)$ under the map e^{-w} has mapping radius exceeding 1, then $f \in R_1$.*

For applications, a variant of the foregoing procedure gives a more useful result. We let Δ be the difference operator defined by

$$(\Delta f)(z) = f(z + 1) - f(z),$$

defining

$$\Delta^0 f = f$$

and

$$\Delta^{n+1} f = \Delta(\Delta^n f).$$

Now, using Taylor's Theorem, we may write

$$\Delta = e^D - 1$$
$$\Delta^n = (e^D - 1)^n,$$

where

$$D = \frac{d}{dz}.$$

We define the functionals T_n and T_n^* by

$$T_n f = f(n)$$
$$T_n^* f = (\Delta^n f)(0).$$

To illustrate,

$$T_0^* = f(0)$$
$$T_1^* f = f(1) - f(0)$$
$$T_2^* f = f(2) - 2f(1) + f(0)$$
$$T_3^* f = f(3) - 3f(2) + 3f(1) - f(0).$$

It is easy to show that

$$T_n^* = (-1)^n \sum_0^n (-1)^k \binom{n}{k} T_k.$$

$$T_n = \sum_0^n \binom{n}{k} T_k^*.$$

For example, the first identity is proved on writing

$$\Delta^n = (e^D - 1)^n = \sum_{k=0}^n \binom{n}{k} e^{kD} (-1)^{n-k}.$$

Definition. To say that a sequence $\{b_n\}$, $n = 0, 1, 2, \ldots$, is K^*-*admissible* is to say that there is an $f \in R$ such that

$$T_n^* f = b_n, \quad n = 0, 1, 2, \ldots.$$

For $f \in R$, write

$$g(z) = \sum_0^\infty (T_n^* f) z^n.$$

It is easily seen that

$$g(z) = \frac{1}{2\pi i} \int_\Gamma \frac{1}{1 - z\zeta(w)} \Phi(d) dw,$$

where $\zeta(w) = e^w - 1$ and Γ is a curve that winds once around $D(f)$. We see that g is analytic outside the image of $D(f)$ under the map $(e^w - 1)^{-1}$. The argument may be reversed to prove the next result.

Theorem. *A sequence $\{b_n\}$ is K^*-admissible if and only if $\sum b_n z^n$ is analytic on the segment $[-1, 0]$.*

We may now prove the main result of this section.

Theorem. *If $f \in R_1$, let $\sum$ be the complement of the image of $D(f)$ under the mapping $(e^w - 1)^{-1}$ and let $\sum^*$ be the image of $D(f)$ under the mapping $e^w - 1$. If $\rho(\sum) > 1$, then $f \in R_1$ and $f(z) = \sum P_k(z)(1 + \beta_k)^z$, where the P_k are polynomials and the β_k run through the complete sets of conjugate algebraic integers lying in $\sum^*$.*

Proof. Let

$$g(z) = \sum (T_k^* f) z^k,$$

and let

$$F(z) = \sum (T_k f) z^k.$$

As we have seen,

$$g(z) = \frac{1}{2\pi i} \int_\Gamma \Phi(w) \frac{1}{1 - z(e^w - 1)} dw$$

$$F(z) = \frac{1}{2\pi i} \int_\Gamma \Phi(w) \frac{1}{1 - z e^w} dw.$$

Now,

$$g(z) = \frac{1}{2\pi i} \frac{1}{1 + z} \int_\Gamma \Phi(w) \frac{1}{1 - \frac{z}{1+z} e^w} dw$$

so that

$$g(z) = \frac{1}{1+z} F\left(\frac{z}{1+z}\right).$$

Similarly,

$$F(w) = \frac{1}{1-w} g\left(\frac{w}{1-w}\right).$$

Since $\rho\left(\sum\right) > 1$, we see by the Pólya Theorem that $g = \frac{P}{Q}$, where P and Q are polynomials with integer coefficients and $Q(0) = 1$. Thus,

$$F(w) = \frac{1}{1-w} \frac{P(\frac{w}{1-w})}{Q(\frac{w}{1-w})} = \frac{(1-w)^N}{(1-w)^{N+1}} \frac{P(\frac{w}{1-w})}{Q(\frac{w}{1-w})} = \frac{P^*(w)}{Q^*(w)},$$

where we choose $N \geq \max(\deg P, \deg Q)$. Now, P^* and Q^* are polynomials with integer coefficients and $Q^*(0) = 1$. Thus, $f \in R_1$. As before, we see that

$$f(z) = \sum P_i(z)\gamma_i^z,$$

where the γ_i are the reciprocals of the roots of Q^* and the P_i are polynomials. If we write $\gamma_i = 1 + \beta_i$, we see that β_i^{-1} is a root of Q, and since the roots of Q are the singularities of g, the theorem is proved.

Using this theorem and some facts about algebraic numbers, the next two results can be proved easily. We state them without proof, as illustrative applications. For details, see the paper of Buck [7].

Theorem. *If f is an integer-valued function of exponential type such that $h_f(\pi/2) = h_f(-\pi/2) = 0$ (that is, the indicator diagram of f is a horizontal line segment), and if $L = \exp h_f(0) - \exp h_f(\pi) < 4$, then $f \in R_1$.*

Theorem. *If, in addition, $L < \sqrt{5}$, then for some polynomials $P_0, P_1, \ldots, P_n$, we have*

$$f(z) = P_0(z) + P_1(z)2^z + \cdots + P_n(z)n^z.$$

22
On Small Entire Functions of Exponential-Type with Given Zeros

This chapter is extracted from a paper of the same name by P. Malliavin and L. A. Rubel [22]. We obtain here a result that considerably generalizes Carlson's Theorem presented in Chapter 20.

For a sequence Λ of positive real numbers, we denote by $F(\Lambda)$ the ideal, in the ring of all entire functions, of those entire functions that vanish at least on Λ. (We exclude once and for all the null function $f = 0$ and the ideal containing only the null function.) We introduce an order relation in this system of ideals, $F(\Lambda) < F(\Lambda')$, meaning that for each $g \in F(\Lambda')$, there is an $f \in F(\Lambda)$ such that $|f(iy)| \leq |g(iy)|$ for every real y. Crudely stated, $F(\Lambda) < F(\Lambda')$ if it is easier to construct small entire functions that vanish on Λ than those that vanish on Λ'.

The major problem is to decide, by elementary computations on Λ and Λ', whether $F(\Lambda) < F(\Lambda')$; we solve this problem here. By specialization, then, we prove as a corollary the following result.

Theorem 22.1. *There exists a function $f \in F(\Lambda)$ such that $|f(iy)| \leq \exp \pi b|y|$ if and only if*

$$\lambda(y) - \lambda(x) \leq b \log \left(\frac{y}{x}\right) + O(1), \quad x \leq y,$$

where $\lambda(t)$ is the sum of the reciprocals of the elements of Λ that do not exceed t.

Remark. Carlson's Theorem deals with the case $b = 1$ and $\Lambda = \{1, 2, 3, \dots\}$. The main innovation of our method is to give our entire functions suitable zeros on the imaginary axis, in addition to the required real zeros.

We proceed now to the body of the exposition. We study sequences $\Lambda = \{\lambda_n\}$ of positive real numbers,

$$\Lambda : 0 < \lambda_0 \le \lambda_1 \le \cdots ,$$

and define

$$\lambda(t) = \sum_{\lambda_n \le t} \lambda_n^{-1}$$

$$\Lambda(t) = \sum_{\lambda_n \le t} 1 = \int_0^t s\, d\lambda(s).$$

Definition. $\lambda(t)$ is called the *characteristic logarithm* of Λ, and $\Lambda(t)$ is called the *counting function* of Λ.

For simplicity, we suppose that Λ is an infinite sequence and that

$$\bar{D}(\Lambda) = \limsup_{t \to \infty} \frac{\Lambda(t)}{t} < \infty,$$

since the problem is trivial if Λ is finite or if $\bar{D}(\Lambda) = \infty$. The function $W(z) = W(z : \Lambda)$ belonging to $F(\Lambda)$ is called the *Weierstrass product* (over Λ) and is defined by

$$W(z : \Lambda) = \Pi \left(1 - \frac{z^2}{\lambda_n^2} \right).$$

We may write

$$\log |W(z : \Lambda)| = \int_0^\infty \log \left| 1 - \frac{r^2}{t^2} e^{2i\theta} \right| d\Lambda(t),$$

where $z = re^{i\theta}$. For $\theta \ne 0, \pi$ we have, on integrating by parts,

$$\log |W(re^{i\theta})| = r \int_0^\infty P(t, \theta) \Lambda(rt) \frac{1}{rt}\, dt,$$

where

$$P(t, \theta) = 2 \frac{1 - t^2 \cos 2\theta}{1 - 2t^2 \cos 2\theta + t^4}.$$

We define, for $0 < b < \infty$, the arithmetic progression Λ_b by

$$\Lambda_b = \left\{ \frac{1}{b}, \frac{2}{b}, \frac{3}{b}, \cdots \right\}$$

and observe that

$$\Lambda_b(t) = [bt] = bt + O(1)$$
$$\lambda_b(t) = b \log t + O(1) \quad (\text{for} \quad t \ge 1)$$

$$W(z : \Lambda_b) = \frac{\sin \pi b z}{\pi b z}$$
$$h_{W_b}(\theta) = \pi b |\sin \theta|.$$

We write $\Lambda \subset \Lambda'$ to indicate that Λ is a subsequence of Λ', and remark that $\Lambda \subset \Lambda'$ if and only if $\lambda(y) - \lambda(x) \le \lambda'(y) - \lambda'(x)$ for $x \le y$.

Definition. Λ is equivalent to Λ', written $\Lambda \sim \Lambda'$, shall mean that $\lambda'(x) - \lambda(x) = O(1)$.

Definition. $\Lambda' > \Lambda$ shall mean that there exists a sequence Λ'', $\Lambda'' \supset \Lambda$, such that $\Lambda'' \sim \Lambda'$.

Definition. $\Lambda < \Lambda'$ shall mean that there exists a sequence Λ''', $\Lambda''' \subset \Lambda'$, such that $\Lambda''' \sim \Lambda$.

Although $\Lambda < \Lambda'$ and $\Lambda' > \Lambda$ mean two different things, the first corollary of the next lemma resolves this notational difficulty.

Lemma 22.2. $\Lambda > \Lambda'$ *if and only if*

$$(22.1) \qquad \lambda(y) - \lambda(x) \leq \lambda'(y) - \lambda'(x) + O(1); \quad 0 < x \leq y < \infty.$$

Likewise, $\Lambda < \Lambda'$ *holds if and only if (22.1) is satisfied.*

Corollary 22.3. $\Lambda < \Lambda'$ *if and only if* $\Lambda' > \Lambda$.

Corollary 22.4. *If* $\Lambda \sim \Lambda_1$, $\Lambda' \sim \Lambda_1'$, *and* $\Lambda < \Lambda'$, *then* $\Lambda_1 < \Lambda_1'$.

Corollary 22.5. *If* $\Lambda_1 < \Lambda_2$ *and* $\Lambda_2 < \Lambda_3$, *then* $\Lambda_1 < \Lambda_3$.

Corollary 22.6. *If* $\Lambda < \Lambda'$ *and* $\Lambda' < \Lambda$, *then* $\Lambda \sim \Lambda'$.

Thus, $<$ is a well-defined partial ordering of equivalence classes under $\sim$.

Proof of Lemma 22.2. That $\Lambda' > \Lambda$ and $\Lambda < \Lambda'$ each imply (22.1) is trivial. To show that (22.1) implies that $\Lambda' > \Lambda$, we define

$$\varphi(x) = \inf\{\lambda'(s) - \lambda(s) : s \geq x\}.$$

It follows from (22.1) that $\varphi(x) \geq -K$ for some constant K.

Now $\varphi(x)$ is constant except for possible jumps at the jumps of $\lambda'(x)$. Let x_0 be a point of discontinuity of φ. Then,

$$\varphi(x_0 - 0) = \lambda'(x_0 - 0) - \lambda(x_0 - 0)$$

and

$$\varphi(x_0 + 0) \leq \lambda'(x_0 + 0) - \lambda(x_0 + 0).$$

We denote by $\Delta\varphi(x_0)$ the jump of φ at x_0. Then

$$(22.2) \qquad \Delta\varphi(x_0) \leq \Delta\lambda'(x_0) - \Delta\lambda(x_0) \leq \Delta\lambda'(x_0).$$

We let

$$\Lambda^*(t) = [\varphi(t)],$$

where $[a]$ denotes the integral part of a and

$$\Phi(t) = \int_0^t s \, d\varphi(s),$$

and let $\lambda^*(t)$ be the characteristic logarithm of that sequence Λ^* whose counting function is $\Lambda^*(t)$. The function $\lambda^*(t)$ is constant except possibly at the jumps of $\varphi(t)$, and we have

$$\Delta\lambda^*(x_0) < \frac{1}{x_0} + \Delta\varphi(x_0).$$

Using (22.2), we get

(22.3) $$\Delta\lambda^*(x_0) < \frac{1}{x_0} + \Delta\lambda'(x_0).$$

Furthermore, $x_0\Delta\lambda^*(x_0)$ and $x_0\Delta\lambda'(x_0)$ must be integers, so that (22.3) implies

(22.4) $$\Delta\lambda^*(x_0) \le \Delta\lambda'(x_0),$$

and this means that Λ^* is a subsequence of Λ'.

We now define

$$\lambda''(x) = \lambda(x) + \lambda^*(x),$$

so that $\lambda''(x)$ is the characteristic logarithm of some sequence $\Lambda'' \supset \Lambda$. To prove that $\Lambda'' \sim \Lambda'$, we must prove that $\delta(x) = O(1)$, where

$$\delta(x) = \lambda(x) + \lambda^*(x) - \lambda'(x).$$

Now,

$$\varphi(t) - \varphi(0) = \int_0^t \frac{1}{s} \, d\Phi(s)$$

and

$$\lambda^*(t) = \int_0^t \frac{1}{s} \, d[\Phi(s)].$$

An integration by parts shows that

$$\lambda^*(t) - \varphi(t) = -\varphi(0) + O\left(\frac{1}{t}\right),$$

so that it is enough to prove that $\theta(x) = O(1)$, where

$$\theta(x) = \lambda(x) + \varphi(x) - \lambda'(x) = \lambda(x) - \lambda'(x) + \inf\{\lambda'(s) - \lambda(s) : s \ge x\}.$$

But it is clear that $\theta(x) \le 0$, and (22.1) is simply another way of saying that $\theta(x) \ge O(1)$.

To prove that (22.1) implies that $\Lambda < \Lambda'$, we put

(22.5) $$\lambda'''(x) = \lambda'(x) - \lambda^*(x).$$

Since by (22.4) Λ^* is a subsequence of Λ', there is a subsequence Λ''' defined by (22.5) and Λ''' is a subsequence of Λ'. Since we already have shown that $\Lambda''' \sim \Lambda$, i.e., $\delta(x) = O(1)$, the proof is complete.

We now state the main result.

Main Theorem. *Given Λ and Λ', the following three statement are equivalent*

(i) $F(\Lambda) < F(\Lambda')$.

(ii) $\Lambda < \Lambda'$.

(iii) *There exists a single pair, f_0, g_0 with $f_0 \in F(\Lambda)$, $g_0 \in F(\Lambda')$, $|f_0(iy)| \le |g_0(iy)|$ for all real y and such that the only zeros of g_0 in the open right half-plane belong to Λ'.*

Theorem 22.1 is a direct corollary of this result. Given Λ and b, choose $\Lambda' = \Lambda_b$ and $g_0(z) = \sin \pi b z$. Since $|g_0(iy)| \sim e^{\pi b |y|}$, the equivalence of (ii) and (iii) proves Theorem 22.1.

Proof of the Main Theorem. We leave the proof that (ii) implies (i) for later. It is clear that (i) implies (iii); a suitable choice for $g_0(z)$ in (iii) is the Weierstrass product $W(z : \Lambda')$. We now prove that (iii) implies (ii). We write f and g instead of f_0 and g_0. Now we choose ρ with $0 < \rho < \lambda_0$ so that all the zeros, $z_n = r_n e^{i\theta_n}$, of f in the right half-plane (assuming for convenience that f has no zeros on $z = iy$) satisfy $r_n > \rho$, and write one form of Carleman's Theorem (Chapter 12), taking $y > x > \rho$ as

$$(22.6) \qquad \Sigma(y) - \Sigma(x) = I(y) - I(x) + J(y) - J(x) + O(1),$$

where

$$\Sigma(R) = \Sigma(R : f) = \sum_{r_n \le R} \left(\frac{1}{r_n} - \frac{r_n}{R^2} \right) \cos \theta_n,$$

$$I(R) = I(R : f) = \frac{1}{2\pi} \int_\rho^R \left(\frac{1}{t^2} - \frac{1}{R^2} \right) \log |f(it)f(-it)| \, dt,$$

$$J(R) = J(R : f) = \frac{1}{\pi R} \int_{-\frac{\pi}{2}}^{\frac{\pi}{2}} \log |f(Re^{i\theta})| \cos \theta \, d\theta.$$

Now

$$(22.7) \qquad \sum \frac{r_n}{R^2} \cos \theta_n = O(1)$$

since

$$\left| \sum \frac{r_n}{R^2} \cos \theta_n \right| \le \sum \frac{r_n}{R^2} = \frac{1}{R} \sum \frac{r_n}{R} \le \frac{1}{R} n(R) = O(1),$$

where $n(r)$ counts the number of zeros of f whose modulus does not exceed r.

Also,

$$(22.8) \qquad J(R) = O(1)$$

since

$$\left| \frac{1}{\pi R} \int_{-\frac{\pi}{2}}^{\frac{\pi}{2}} \log |f(Re^{i\theta})| \cos\theta d\theta \right| \leq \frac{1}{\pi R} \int_{-\pi}^{\pi} |\log |f(Re^{i\theta})|| \, d\theta$$

$$\leq \frac{1}{\pi R} O(R) = O(1),$$

since

$$\frac{1}{2\pi} \int_{-\pi}^{\pi} |\log |f(Re^{i\theta})|| \, d\theta = m(R, f) + m\left(R, \frac{1}{f}\right) \leq 2T(R, f)$$

and f is of exponential-type.

From (22.7), we obtain

(22.9) $$\Sigma(y) - \Sigma(x) \geq \lambda(y) - \lambda(x) + O(1),$$

and using (22.8) and (22.9) in (22.6) we get

(22.10) $$\lambda(y) - \lambda(x) \leq I(y) - I(x) + O(1).$$

Now, since $|f(iy)| \leq |g(iy)|$, we see that

(22.11) $$I(y : f) - I(x : f) \leq I(y : g) - I(x : g).$$

On the other hand, applying Carleman's Theorem now to g, whose only zeros in the right half-plane are the λ'_n, we see that

(22.12) $$I(y : g) - I(x : g) = \lambda'(y) - \lambda'(x) + O(1).$$

Combining (22.12) with (22.11) and (22.10), we get

$$\lambda(y) - \lambda(x) \leq \lambda'(y) - \lambda'(x) + O(1),$$

and the proof is complete.

To prove now that (ii) implies (i), we suppose that $\lambda(y) - \lambda(x) \leq \lambda'(y) - \lambda'(x) + O(1)$, and we are given $g \in F(\Lambda')$; we must construct a function $f \in F(\Lambda)$ with $|f(iy)| \leq |g(iy)|$ for all y. By Lemma 23.1, we may suppose that $\lambda(t) = \lambda'(t) + O(1)$, since Λ is a subsequence of a sequence Λ'' for which this is true, and $F(\Lambda'') \subset F(\Lambda)$.

By the Hadamard Factorization Theorem we may write

$$g(z) = g_1(z)g_2(z),$$

where

$$g_1(z) = \Pi \left(1 - \frac{z}{\lambda'_n}\right) \exp\left(\frac{z}{\lambda'_n}\right),$$

$$g_2(z) = Cz^k e^{az} \Pi \left(1 - \frac{z}{\zeta_n}\right) \exp\left(\frac{z}{\zeta_n}\right),$$

where the $\zeta_n \neq 0$ are the zeros of g that are not counted in Λ'.

Writing $\log |g_1(iy)|$ as a sum of logarithms, and that sum as a Stieltjes integral, we get

$$(22.13) \qquad \log |g_1(iy)| = \frac{1}{2} \int_0^\infty \log \left(1 + \frac{y^2}{t^2}\right) t \, d\lambda'(t).$$

The next lemma provides the main tool of our construction; it will enable us to "move the zeros" from the real axis to the imaginary axis.

Lemma 22.7. *Let $d\Delta$ be a measure with compact support contained in an interval $[\epsilon, \epsilon^{-1}]$ for some small $\epsilon > 0$. Then there exists a function $\varphi(t)$ defined on $(0, \infty)$ such that*

$$(22.14) \qquad \int \log \left(1 + \frac{y^2}{t^2}\right) d\Delta(t) = \int \log \left|1 - \frac{y^2}{t^2}\right| \varphi(t) \, d\Delta(t)$$

and

$$|\varphi| < 2 \sup_x \left| \int_0^x \frac{1}{s} \, d\Delta(s) \right|.$$

Proof. By a contour integration, it is easy to see that

$$\log |1 + x^2| = \frac{2}{\pi} \int_0^\infty \log \left|1 - \frac{x^2}{t^2}\right| \frac{t}{t^2 + 1} \frac{dt}{t}.$$

By Fubini's theorem,
$$(22.15)$$
$$\int \log \left(1 + \frac{x^2}{r^2}\right) d\Delta(r) = \frac{2}{\pi} \int \log \left|1 - \frac{x^2}{w^2}\right| \left\{ \int \frac{w/t}{w^2/t^2 + 1} \, d\Delta(t) \right\} \frac{dw}{w}.$$

We therefore are led to define

$$(22.16) \qquad \varphi(w) = \frac{2}{\pi} \int \frac{t^2}{w^2 + t^2} \frac{d\Delta(t)}{t},$$

and (22.14) asserts (22.13) in another form. The bound on φ follows from integrating by parts in (22.15):

$$(22.17) \qquad \varphi(w) = \frac{2}{\pi} \int_0^\infty \frac{d\Delta(t)}{t} - \frac{2}{\pi} \int_0^\infty \left\{ \int_0^x \frac{d\Delta(t)}{t} \right\} d_x \left(\frac{x^2}{x^2 + w^2}\right).$$

Hence,

$$|\varphi(w)| < \left\{1 + \int_0^\infty d_x \left(\frac{x^2}{x^2 + w^2}\right)\right\} \sup_x \left| \int_0^x \frac{d\Delta(t)}{t} \right|$$

since $\frac{x^2}{x^2+w^2}$ is increasing, and the lemma is proved.

We now choose

(22.18) $$\delta(t) = \frac{1}{2}\{\lambda'(t) - \lambda(t)\}, \quad d\Delta(t) = t \, d\delta(t)$$

but cannot apply Lemma 22.7 to $d\Delta$ since its support may not be compact. We truncate the support by defining

$$\delta_k(t) = \begin{cases} \delta(t) & \text{if } t \le k \\ \delta(k) & \text{if } t > k \end{cases}$$
$$\Delta_k(t) = t \, d\delta_k(t)$$

with the same convention for $\lambda(t)$ and $\lambda'(t)$.

We now apply Lemma 22.7 to $d\Delta_k$ and conclude that there exist functions $\varphi_k(t)$ such that

(22.19) $$\int \log\left(1 + \frac{y^2}{t^2}\right) d\Delta_k(t) = \int \log\left|1 - \frac{y^2}{t^2}\right| \varphi_k(t) \, dt.$$

Now,

(22.20) $$|\varphi_k(t)| \le B,$$

where B is a constant that is independent of k, namely, from the bound on $|\varphi(t)|$ in the lemma and the equivalence of Λ and Λ',

$$B = 2 \sup |\lambda(t) - \lambda'(t)|.$$

On putting

(22.21) $$L_k(y) = \int \log\left(1 + \frac{y^2}{t^2}\right) \frac{1}{2} t \, d\lambda_k(t) + \int \log\left|1 - \frac{y^2}{t^2}\right| d\Phi_k(t),$$

where

(22.22) $$d\Phi_k(t) = \varphi_k(t) \, dt,$$

we have

(22.23) $$L_k(y) = \frac{1}{2} \int \log\left(1 + \frac{y^2}{t^2}\right) t \, d\lambda'_k(t).$$

Hence, by (22.13),

(22.24) $$\lim_{k\to\infty} L_k(y) = \log|g_1(iy)|.$$

At this point, the idea is to find an entire function F for which the hypothetical formula

$$\log |F(iy)| = \lim_{k \to \infty} \int \log \left|1 - \frac{y^2}{t^2}\right| d\Phi_k(t)$$

holds in some appropriate sense. First, however, the limit need not exist, but a simple argument with normal families will handle this difficulty. Also, the measures $d\Phi_k(t) = \varphi_k(t)dt$ are unsuitable since they need not be positive and cannot be discrete. [It is easy to see that all the $d\Phi_k(t)$ are positive only in case $\Lambda \subseteq \Lambda'$, a trivial case.] But first we show that adding a constant to φ_k, in order to make $d\Phi_k(t)$ positive, does not change L_k. Then we show that the resulting measure may be made discrete with little loss of precision.

Resuming the construction, we define

$$(22.25) \qquad\qquad \psi_k(t) = B + \varphi_k(t)$$

and by (22.20) conclude that

$$(22.26) \qquad\qquad \psi_k(t) \geq 0 \quad \text{for all} \quad t.$$

A contour integration shows that

$$(22.27) \qquad\qquad \int \log \left|1 - \frac{y^2}{t^2}\right| dt = 0,$$

so that

$$(22.28) \qquad L_k(y) = \int \log \left(1 + \frac{y^2}{t^2}\right) \frac{t}{2} \, d\lambda_k(t) + \int \log \left|1 - \frac{y^2}{t^2}\right| d\Psi_k(t),$$

where $d\Psi_k(t) = \psi_k(t) \, dt$. Now let $\Psi_k^*(t) = [\Psi_k(t)]$, the integral part of $\Psi_k(t)$, and define

$$(22.29) \qquad L_k^*(y) = \int \log \left(1 + \frac{y^2}{t^2}\right) \frac{t}{2} \, d\lambda_k(t) + \int \log \left|1 - \frac{y^2}{t^2}\right| d\Psi_k^*(t).$$

Lemma 22.8. *There is a constant β, independent of k, such that for all $y > 1$*

$$(22.30) \qquad \int \log \left|1 - \frac{y^2}{t^2}\right| d\Psi_k^*(t) \leq \int \log \left|1 - \frac{y^2}{t^2}\right| d\Psi_k(t) + \beta \log |y|.$$

Proof. We apply the next lemma with $\Psi_k(t) = \nu(t)$ and $\Psi_k^*(t) = n(t)$. β is independent of k because $\left|\frac{d}{dt}\Psi_k(t)\right|$ and $\left|\Psi_k(t) - \Psi_k^*(t)\right|$ are bounded independently of k.

Lemma 22.9. *Suppose that $\nu(r)$ is a continuously differentiable function for $0 < r < \infty$, that $0 \le \nu'(r) < B < \infty$, that $n(r)$ is nondecreasing, and that for some constant C*

$$\nu(r) \ge n(r) > \nu(r) - C.$$

Then

$$\int \log \left|1 - \frac{y^2}{t^2}\right| \, dn(t) \le \int \log \left|1 - \frac{y^2}{t^2}\right| \, d\nu(t) + O(\log y)$$

as $y \to \infty$.

Proof. For fixed r, we write $L(t) = \log\left|1 - \frac{r^2}{t^2}\right|$ and point out that L is Lebesgue integrable on $(0, \infty)$:

$$L(0^+) = +\infty, \quad L(r^-) = L(r^+) = -\infty, \quad L(\infty) = 0$$

and that $L(t)$ is decreasing and continuous for $t \in (0, r)$ and increasing and continuous for $t \in (r, \infty)$. We must compare $Y = \int_0^\infty L(t) dn(t)$ and $Z = \int_0^\infty L(t) d\nu(t)$. We will prove that $Y < Z + O(\log r)$. We assume that $\nu'(t) \ge p > 0$. This involves no loss of generality since, if we replace $\nu(t)$ by $\nu(t) + t$ and $n(t)$ by $n(t) + t$, we change Y and Z not at all, because $\int_0^\infty L(t) dt = 0$. We may suppose, without loss of generality, that $\nu(0) = 0$, since suitably redefining ν on the interval $[0, 1]$ changes Z only by $O(1)$, which is small compared to the allowed discrepancy $O(\log r)$.

With each large r we associate the numbers r_1 and r_2 such that

$$\nu(r_1) = n(r) = \nu(r_2) - C.$$

Since $\nu'(t) \ge p$, we will have $r - r_1 \le r_2 - r_1 \le \frac{C}{p}$. It is easy to see that the following inequalities hold:

$$\int_0^r L(t) \, dn(t) \le \int_0^{r_1} L(t) \, d\nu(t),$$

$$\int_r^\infty L(t) \, dn(t) \le \int_{r_2}^\infty L(t) \, d\nu(t).$$

It follows that $Y = X + Z$, where

$$X = -\int_{r_1}^{r_2} \log \left|1 - \frac{y^2}{t^2}\right| \, d\nu(t).$$

We shall prove that $X \le O(\log r)$. Clearly,

$$X \le -\int_{r_1}^{r_2} \log \left|\frac{t - r}{t}\right| \, d\nu(t).$$

Since $r_2 - r_1 \leq \frac{C}{p}$ and $\nu'(t) \leq B$, we have

$$X \leq -B \int_{r_1}^{r_2} \log^- \left| \frac{t-r}{t} \right| \, dt \leq B(r_2 - r_1) \log r_2 - B \int_{r_1}^{r_2} \log^- |t - r| \, dt$$

so that

$$X \leq \frac{BC}{p} \log \left(r + \frac{C}{p} \right) + 2B.$$

We now consider polynomials P_k defined by

$$\log |P_k(z)| = \int \log \left| 1 + \frac{z^2}{t^2} \right| \, d\Psi_k^*(t).$$

Lemma 22.10. *There is a constant B', independent of k, such that for all z*

(22.31) $$\log |P_k(z)| \leq B'|z|.$$

Proof. Putting $z = x + iy$, we see that

$$\int \log \left| 1 + \frac{z^2}{t^2} \right| \, d\Psi_k^*(t) \leq \int \log \left| 1 + \frac{x^2}{t^2} \right| \, d\Psi_k^*(t).$$

But since $\Psi_k^*(t) = \left[\int_0^t \{B + \varphi_k(s)\} ds \right]$, and since $|\varphi_k(s)| \leq B$ by (22.20), the proof is immediate on integrating by parts.

Since the family $\{P_k(z)\}$ is therefore uniformly bounded in each disk $D_R = \{z : |z| < R\}$, it is consequently a normal family, and we may extract a subsequence $\{P_{k'}(z)\}$ that converges uniformly on compact sets to an entire function F:

(22.32) $$F(z) = \lim_{k' \to \infty} P_{k'}(z).$$

Because $P_k(0) = 1$ for each k, it follows that $F(0) = 1$. By Lemma 22.10, we conclude that F is of exponential-type. Since P_k has only imaginary zeros, so has F. Furthermore, F is an even function since each P_k is.

Let $i\Gamma = \{i\gamma_n\}$ be the zeros of F on the positive imaginary axis. Since F is of exponential-type, Γ has finite upper density. Thus,

(22.33) $$F(z) = \Pi \left(1 + \frac{z^2}{\gamma_n^2} \right).$$

At last we can define $f(z)$. Let

(22.34) $$f(z) = f_1(z) F(z) g_2(z),$$

where

$$(22.35) \qquad f_1(z) = \Pi \left(1 - \frac{z}{\lambda_n}\right) \exp\left(\frac{z}{\lambda_n}\right).$$

As a consequence of the estimates (22.24) and (22.30) we see that

$$(22.36) \qquad \log |f(iy)| \le \log |g(iy)| + O(\log |y|).$$

Now (22.36) is as good for our purposes as $|f(iy)| \le |g(iy)|$, since we could otherwise consider $f^*(z) = f(z)a\{(iz)^{-1}\sin(iz)\}^b$ for a suitable choice of a and b.

It is not obvious, though, that f is of exponential-type, since f_1 and g_2 need not be of exponential-type, although they are certainly of order 1. To prove that f is of exponential-type we appeal to Lindelöf's Theorem proven in Chapter 13.

Let us denote by a_n and b_n the zeros, other than the origin, of f and g, respectively. Then we see that

$$O(1) = \sum_{|b_n| \le R} b_n^{-1} = \sum_{\substack{|b_n| \le R \\ b_n \notin \Lambda'}} b_n^{-1} + \lambda'(R)$$

$$= \sum_{\substack{|b_n| \le R \\ b_n \notin \Lambda'}} b_n^{-1} + \lambda(R) + (\lambda'(R) - \lambda(R))$$

$$= \sum_{\substack{|b_n| \le R \\ b_n \notin \Lambda'}} b_n^{-1} + \lambda(R) + O(1)$$

since $\lambda'(R) - \lambda(R) = O(1)$ by hypothesis. Now the zeros of f, other than the origin, fall into three categories: (i) those b_n not counted in Λ', (ii) the elements of Λ, and (iii) the zeros of F. If we consider

$$S(R) = \sum_{|b_n| \le R} b_n^{-1},$$

the zeros of F contribute nothing to $S(R)$ since F is even. Hence,

$$S(R) = \sum_{\substack{|b_n| \le R \\ b_n \notin \Lambda'}} b_n^{-1},$$

and it follows that

$$S(R) = O(1).$$

This, together with the obvious fact that the zeros of f have finite upper density, implies (via Lindelöf's Theorem) that f is of exponential-type. The proof is complete.

23
The First-Order Theory of the Ring of All Entire Functions

The material of this chapter is drawn from the paper [3], "First-Order Conformal Invariants." Let $\mathcal{E}$ denote the ring of all entire functions as an abstract ring. Much information about the theory of entire functions is present in the theory of $\mathcal{E}$. For example, an entire function f omits the value 7 iff there exists an entire function g such that $(f - 7)g = 1$.

We show that even using a restricted logic (first-order logic), a great deal can be expressed in the theory of $\mathcal{E}$. We shall show, indeed, that *all* of classical function theory can be so expressed. By the *ring language* we mean the first-order formal language appropriate to the structure $\mathcal{E}$. This language has basic symbols for addition and multiplication of entire functions, as well as the usual logical symbols: $\wedge$ ("and"), $\vee$ ("or"), $\neg$ ("not") and $\Longrightarrow$ ("implies") as well as quantifier symbols $\forall$ ("for all") and $\exists$ ("there exists") together with variables that range over $\mathcal{E}$. For convenience we also include in the ring language a constant symbol which is a name for the constant function $i = \sqrt{-1}$. (Otherwise, there would be no way to distinguish between the two solutions of $f^2 + 1 = 0$.) Formulas and sentences in this language are finite combinations of these basic symbols, arranged according to the obvious formal rules of grammar. A key restriction is that the language is *first-order*, which means that we can only use quantifiers over *elements* of $\mathcal{E}$ and not over subsets, ideals, relations, etc. Also, the expressions in a first-order language are always finite in length. (See [39] for a general treatment.)

The *algebra language* is appropriate to $\mathcal{E}$ as an algebra. This is formed by adding to the ring language a 1-place predicate Const. In $\mathcal{E}$ (as an algebra) we interpret $\text{Const}(f)$ to mean that f is a constant function. (Thus we are identifying $\mathbb{C}$ with the field of constant functions in $\mathcal{E}$ and are using the ordinary addition and multiplication of functions in $\mathcal{E}$ to play the role of addition of constants and the scalar multiplication in the algebra.)

But in $\mathcal{E}$ it is easy to say that f is a constant (either $f = 0$, $f = 1$, or f omits 0 and 1) so that the ring language and the algebra language are equivalent.

In dealing with rings and algebras, the expressive power of first-order sentences is reasonably well understood. For example, to say that a ring is commutative is first-order $(\forall x \forall y (xy = yx))$ but, at least superficially, to say that a ring is simple is not first-order, since it seems to require quantification over subsets (there does not exist a proper two-sided ideal), and in fact does so require. It is first-order to say that a ring has at least two elements $(\exists x \exists y (x \neq y))$, but it is not first-order to say that a ring is infinite, since this requires a sentence of infinite length, as one might suspect.

Nearly all of the results in this chapter depend on the fundamental definability results for the algebra $\mathcal{E}$. We show that there are formulas in the algebra language which define in $\mathcal{E}$ the set $\mathbb{N}$ of nonnegative integers, the set $\mathbb{Z}$ of all the integers, the field of rational numbers $\mathbb{Q}$, the field of real numbers $\mathbb{R}$, the ordering relation $\leq$ on $\mathbb{R}$, and the absolute value function on $\mathbb{C}$. We also show how to interpret in the first-order language of $\mathcal{E}$ the quantifiers that range over countable sets and sequences of constants. (It is striking that second-order concepts can be represented within the restricted first-order language.)

An immediate consequence of this is that second-order number theory is interpretable in the first-order theory of $\mathcal{E}$. Other recursive undecidability results are treated later. For example, we show that the first-order theory of the ring of entire functions is recursively isomorphic to second-order number theory. (This improves on a result of Robinson [33], who showed how to interpret first-order number theory in the first-order theory of entire functions.)

Later we extend these definability results even further. We give enough examples to suggest that all of classical analytic function theory on $\mathbb{C}$ can be interpreted in the first-order theory of $\mathcal{E}$. This is quite surprising, given the apparent limitations of the first-order algebra language. We also show how to interpret in the first-order language of $\mathcal{E}$ the quantifiers that range over countable subsets and sequences in $\mathcal{E}$ itself. That is, the first-order language is already as expressive in this context as the restricted second-order language.

The Ring Language
In this section we will begin to explore the expressive power of the first-order theory of the ring $\mathcal{E}$. To begin, note that the constant functions 0 and 1 are definable using their first-order properties in $\mathcal{E}$. Also, the property that f is a unit in $\mathcal{E}$ is first-order expressible: $\exists g (fg = 1)$. Thus we can

express, for any definable constant c, the condition that f omits the value c on $\mathbb{C}$ by saying that $f - c$ is a unit in $\mathcal{E}$. Since 1 and i are definable in $\mathcal{E}$, so is each Gaussian rational number. This means that for each Gaussian rational q there is a formula $F_q(x)$ in the algebra language such that, for any function f in $\mathcal{E}$, $F_q(f)$ holds in $\mathcal{E}$ if and only if f equals the constant function q.

In this section we present a detailed study of certain basic definability questions for the algebra $\mathcal{E}$. These matters are fundamental to the general content of this chapter and are of interest in their own right.

We call a function $f \in \mathcal{E}$ a *point function* if it has a unique zero on $\mathbb{C}$, of multiplicity one. These will be used to represent the points of $\mathbb{C}$ within $\mathcal{E}$. It is easy to construct a formula $P(x)$ in the algebra language such that $P(f)$ holds in $\mathcal{E}$ if and only if f is a point function. That is, $P(f)$ should express

$$f \quad \text{is not a unit, and}$$
$$\forall g \forall h [f = gh \implies g \quad \text{is a unit or } h \text{ is a unit}].$$

Using this we can, for example, construct an algebra language formula $A(x)$ such that $A(f)$ holds in $\mathcal{E}$ if and only if f is a $1-1$ conformal mapping on $\mathbb{C}$. That is, $A(f)$ should express the condition:

For any constant α and any point functions g and h,

if both g and h divide $f - \alpha$, then g divides h.

[To say that a point function g divides $f - \alpha$ is equivalent to saying that f has value α at the unique point of $\mathbb{C}$ where g is zero. To say that one point function divides another is equivalent to asserting that they are zero at the same point of $\mathbb{C}$. Of course the conformal maps of $\mathbb{C}$ are exactly the functions $f(z) = az + b$, $a \neq 0$.]

Next we discuss how to code arbitrary countable or finite sets of constants in $\mathbb{C}$ using first-order formulas. It is very striking that we can represent second-order mathematical concepts in a first-order language.

Given $f, g \in \mathcal{E}$, define $V(\alpha; f, g)$ to mean that α is a constant, $g \neq 0$, and there exists $z_0 \in \mathbb{C}$ such that $g(z_0) = 0$ and $f(z_0) = \alpha$. This can be represented by an algebra language formula, since $V(\alpha; f, g)$ is equivalent to the existence of a point function h such that h divides g and h divides $f - \alpha$ (together with the other conditions, that α is constant and $g \neq 0$). We think of the pair (f, g) as coding the set of constants

$$E = \{\alpha \in \mathbb{C} \mid V(\alpha; f, g) \text{ holds in } \mathcal{E}\}.$$

If $g = 0$, then this set is empty. If $g \neq 0$, then it is a countable or finite subset of $\mathbb{C}$. Moreover, by taking g to have an infinite sequence of zeros tending to ∞ and by letting f vary over $\mathcal{E}$, then we obtain for E all possible countable or finite subsets of $\mathbb{C}$ [38].

We may use this idea, for example, to find a formula $IZ(x)$ in the algebra language such that $IZ(g)$ holds in $\mathcal{E}$ if and only if the zero set of g in $\mathbb{C}$ is infinite while $g \neq 0$. Namely, let $IZ(g)$ be the formula:

$$g \neq 0 \wedge \exists f[V(0; f, g) \wedge \forall \alpha(V(\alpha; f, g) \Longrightarrow V(\alpha + 1; f, g))].$$

This condition asserts the existence of a function f such that the set $E \supseteq \mathbb{N}$. By the well-known interpolation theorem for entire functions this can happen exactly when g has infinitely many zeros.

Theorem 23.1. *The following sets and relations are all definable in $\mathcal{E}$ by formulas in the algebra language:*

The set $\mathbb{Z}$ of (positive and negative) integers,
That set $\mathbb{N}$ of nonnegative integers,
The set $\mathbb{Q}$ of rational numbers,
The set $\mathbb{Q}(i)$ of Gaussian rationals,
The ordering relations $\leq$ on $\mathbb{Z}$, $\mathbb{N}$, $\mathbb{Q}$, and
The absolute value function on $\mathbb{Q}$.

Proof. This is immediate from the discussion above. To get $\mathbb{N}$ we use the Peano axioms: $\mathbb{N}$ is the smallest countable subset of $\mathbb{C}$ that contains 0 and contains $\alpha + 1$ whenever it contains α. Precisely, the formula $N(\alpha)$ defining $\mathbb{N}$ is

$$\alpha \text{ is constant } \wedge$$
$$\forall f \forall g[\{V(0; f, g) \wedge \forall \beta(V(\beta; f, g)) \Longrightarrow V(\beta + 1, f, g)\}$$
$$\Longrightarrow V(\alpha; f, g)].$$

Once having defined $\mathbb{N}$, $\mathbb{Z}$ and $\mathbb{Q}$ are trivial:

$$\alpha \in \mathbb{Z} \Longleftrightarrow \alpha \in \mathbb{N} \vee -\alpha \in \mathbb{N},$$
$$\alpha \in \mathbb{Q} \Longleftrightarrow \exists \beta \exists \gamma[\beta \in \mathbb{Z} \wedge \gamma \in \mathbb{N} \wedge \gamma \neq 0 \wedge \alpha \gamma = \beta].$$

To define the Gaussian rationals $\mathbb{Q}(i)$ we note that

$$\alpha \in \mathbb{Q}(i) \Longleftrightarrow \exists \beta \exists \gamma[\beta \in \mathbb{Q} \wedge \gamma \in \mathbb{Q} \wedge \alpha = \beta + \gamma i].$$

To define $\leq$, on $\mathbb{Q}$ say, note that

$$\alpha \in \beta \Longleftrightarrow \exists \gamma \exists \delta[\gamma \in \mathbb{N} \wedge \delta \in \mathbb{N} \wedge \gamma \neq 0 \wedge \gamma(\beta - \alpha) = \delta].$$

Finally, for $\alpha \in \mathbb{Q}$,

$$|\alpha| = \beta \Longleftrightarrow 0 \leq \beta \wedge (\beta = \alpha \vee \beta = -\alpha).$$

Note that since $\mathbb{N}$ is definable in $\mathcal{E}$, there is an effective procedure for interpreting the first-order theory of $(\mathbb{N}, +, \cdot)$ into the first-order theory of

$\mathcal{E}$. (The operations $+$ and $\cdot$ are just the natural operations on $\mathcal{E}$.) Actually, this interpretation extends to the second-order theory of $(\mathbb{N}, +, \cdot)$, since we have a way to discuss countable sets of constants in the first-order language of $\mathcal{E}$. It follows that the first-order theory of $\mathcal{E}$ is very undecidable in the sense of recursive function theory. This will be discussed fully below.

We caution the reader that at this stage we are only able to define the field of *rational* constants in $\mathcal{E}$. It is possible to define the real field $\mathbb{R}$, as we show below, but there does not seem to be any easy way to do it.

Once we have a first-order definition of $\mathbb{R}$ in $\mathcal{E}$, no matter what it is, we get immediately as a bonus a first-order definition of $\leqq$ on $\mathbb{R}$ and of the absolute value on $\mathbb{C}$:

$$\alpha \leqq \beta \Longleftrightarrow \exists \gamma [\gamma \in \mathbb{R} \wedge \beta = \alpha + \gamma^2],$$
$$|\alpha| = \beta \Longleftrightarrow \exists \gamma \exists \delta [\gamma \in \mathbb{R} \wedge \delta \in \mathbb{R} \wedge \alpha = \gamma + \delta i$$
$$\wedge \beta \in \mathbb{R} \wedge 0 \leqq \beta \wedge \beta^2 = \gamma^2 + \delta^2].$$

We now turn to the construction of an algebra language formula $R(x)$ that defines the field of real constants in $\mathcal{E}$. The construction of $R(x)$ is complicated, but it is based on an elementary fact: A constant α is a real number if and only if there is a Cauchy sequence S of rationals such that every analytic function that carries S to an equivalent Cauchy sequence of rationals also preserves α. Our problem has two parts: to prove that this equivalence is correct and to construct an algebra language formula that expresses the right side of it. To prove the equivalence we need the following fact:

Lemma A. *If $\beta \in \mathbb{R}$ and $\alpha \in \mathbb{C}\backslash\mathbb{R}$, then there is an entire function f such that $f(\beta) = \beta$, $f(\alpha) \neq \alpha$, and f maps $\mathbb{Q}$ onto $\mathbb{Q}$.*

Proof. This argument is a simple modification of the proof given in [27]. Fix $\beta \in \mathbb{R}$ and $\alpha \in \mathbb{C}\backslash\mathbb{R}$. If $\beta \in \mathbb{Q}$, then f is trivial to find, so we may assume $\beta \notin \mathbb{Q}$ also. We take $p(z) = e^{|z|}$, so that $p(z) \geq e|z|$ for all $z \in \mathbb{C}$. Take $S = T = \mathbb{Q} \cup \{\beta\}$ and choose $f_0 \in EM$ so that

$$f_0(\beta) = \frac{1}{2}\beta, |\Im(f_0(\alpha) - \alpha)| \geq 2p(|\alpha|),$$

and let $\delta = \frac{1}{2}$. (Here, EM is the set of all entire functions whose restriction to $\mathbb{R}$ is a real, monotonically nondecreasing function.) Let us take an enumeration of S (and T) with $x_1 = \beta$ (in particular, $x_1 \neq 0$), and let $\delta > 0$ be such that

$$f_0(x_1) + \delta x_1 \in T \quad \text{and} \quad |\delta z| \leq \frac{1}{2}p(|z|) \quad \text{for all} \quad z \in \mathbb{C},$$

where f_0 is any function chosen from EM. We define

$$f_1(z) = f_0(z) + \delta z, \quad S_1 = \{x_1\} \quad \text{and} \quad T_1 = \{f_1(x_1)\}.$$

Note that $S_1 = \{\beta\}$ and $f_1(\beta) = \beta$, so that $T_1 = \{\beta\}$ also. We now construct the sequences $\{f_n\}$, $\{S_n\}$, and $\{T_n\}$ so that

$$f'_n(x) \geq (2^{-1} + 2^{-n})\delta \quad \text{and} \quad f_n(S_n) = T_n$$

for all $n \in \mathbb{N}$, $x \in \mathbb{R}$. Suppose that f_n, S_n, and T_n have been constructed and choose a polynomial g with real coefficients such that
  (1) $g(z) = 0 \Longleftrightarrow z \in S_n \quad (z \in \mathbb{C})$
  (2) $|g(z)| \leq 2^{-n-1}p(|z|) \quad (z \in \mathbb{C})$
  (3) $g'(x) \geq -2^{-n-1}\delta \quad (x \in \mathbb{R})$.
[Any polynomial of odd degree with positive leading coefficient for which (1) is valid will also obey (2) and (3) after it has been multiplied by a small enough positive constant. The degree can be chosen odd by adjusting the multiplicity of one of the zeros.]
  For each $M \in [0, 1]$ we have

$$(f_n + Mg)'(x) = f'_n(x) + Mg'(x) \geq (2^{-1} - 2^{-n-1})\delta \quad (x \in \mathbb{R}),$$

so that $f_n + Mg$ is strictly increasing on $\mathbb{R}$.
  Moreover, if we let M vary in $[0, 1]$, then for $x \notin S_n$, $(f_n + Mg)(x)$ varies in an interval of $\mathbb{R}$ that contains points of $T \backslash T_n$; and for $y \notin T_n$, $(f_n + Mg)^{-1}(y)$ varies in an interval of $\mathbb{R}$ that contains points of $S \backslash S_n$.
  Now for n odd, let x be the point of $S \backslash S_n$ with smallest index and let $M \in [0, 1]$ be such that $(f_n + Mg)(x) \in T$. We define

$$f_{n+1} = f_n + Mg, \quad S_{n+1} = S_n \cup \{x\}$$

and

$$T_{n+1} = T_n \cup \{f_{n+1}(x)\}.$$

  For n even, let y be the point of $T \backslash T_n$ with smallest index and let $M \in [0, 1]$ be such that $\{f_n + Mg\}^{-1}(y) \in S$. We define

$$f_{n+1} = f_n + Mg, \quad T_{n+1} = T_n \cup \{y\}$$

and

$$S_{n+1} = S_n \cup \{f_{n+1}^{-1}(y)\}.$$

The following properties of the constructed sequences are easily verified:
  (a) $|f_n(z) - f_{n-1}(z)| \leq 2^{-n}p(|z|) \quad (n \in \mathbb{N}, z \in \mathbb{C})$
  (b) $\bigcup_{n=1}^{\infty} S_n = S$, $\bigcup_{n=1}^{\infty} T_n = T$, and

$$f_m(S_n) = T_n, \quad (m, n \in \mathbb{N}, m \geq n).$$

From (a) it follows that f_n converges pointwise to a function f for which

$$|f(z) - f_0(z)| \leq p(|z|) \quad (z \in \mathbb{C}).$$

Now $p(|z|)$ is a function of z that is bounded on compact subsets of $\mathbb{C}$, so the convergence of $\{f_n(z)\}$ is uniform on such sets. From this we conclude that f is an entire function.

For each $n \in \mathbb{N}$ we have $f'_n(x) \geq \frac{1}{2}\delta$, $(x \in \mathbb{R})$, so the same is true for f. Hence, f is strictly increasing on $\mathbb{R}$. From (b), it follows that $f(S_n) = T_n$ for each n, and so $f(S) = T$. Moreover, we have insured that $f_n(\beta) = \beta$ for all $n \in \mathbb{N}$ and therefore $f(\beta) = \beta$. This implies $f(\mathbb{Q}) = \mathbb{Q}$ also.

Finally we have

$$\begin{aligned}|\Im(f(\alpha) - \alpha)| &\geq |\Im(f_0(\alpha) - \alpha)| - |\Im(f_0(\alpha) - f(\alpha))| \\ &\geq 2p(|\alpha|) - p(|\alpha|) > 0,\end{aligned}$$

which means $f(\alpha) \neq \alpha$. This completes the proof.

Now it is easy, using Lemma A, to show that our characterization of the real numbers is correct. In one direction, the equivalence is trivial: If $\alpha \in \mathbb{R}$, we need only take S to be a Cauchy sequence of rationals that converges to α. For the converse direction, suppose that $\alpha \in \mathbb{C}\backslash\mathbb{R}$ and S is any Cauchy sequence from $\mathbb{Q}$. Let β be the limit of S in $\mathbb{R}$. Use Lemma A to get an entire function f such that $f(\beta) = \beta$, $f(\alpha) \neq \alpha$, and $f(\mathbb{Q}) \subseteq \mathbb{Q}$. Then f maps S to a Cauchy sequence of rationals that is equivalent to S (since both converge to β) yet f does not preserve α. That is, $\alpha \in \mathbb{C}\backslash\mathbb{R}$ implies that the right side of our equivalence is false. This completes the proof that our characterization of $\mathbb{R}$ is correct. Now we must express it formally within the algebra language.

Theorem 23.2.   *There are formulas $R(x)$, $L(x,y)$, and $M(x,y)$ in the algebra language such that for any α, $\beta \in \mathcal{E}$:*

(i) $\alpha \in \mathbb{R} \Longleftrightarrow R(\alpha)$ holds in $\mathcal{E}$;

(ii) For α, $\beta \in \mathbb{R}$, $\alpha \leq \beta \Longleftrightarrow L(\alpha,\beta)$ holds in $\mathcal{E}$;

(iii) For α, $\beta \in \mathbb{C}$, $|\alpha| = \beta \Longleftrightarrow M(\alpha,\beta)$ holds in $\mathcal{E}$.

Proof. We begin by building some machinery for discussing *sequences* of constants within the first-order language of $\mathcal{E}$. (Earlier we did the same for countable *sets* of constants.) This is done using a triple of functions (f,g,h); g has infinitely many zeros on $\mathbb{C}$, and on that zero set h takes on the values $0, 1, 2, \ldots$. In effect, h *lists* the zeros of g. Then the sequence coded by (f,g,h) is (α_n), where α_n is the value $f(z_n)$ at the zero of g where h take the value n. First let Basis (g,h) be a formula in the algebra language which expresses the fact that h "lists" the zeros of g in the manner discussed above:

$$\begin{aligned}\text{Basis}(g,h) &\Longleftrightarrow \forall\alpha[V(\alpha; h, g) \Longleftrightarrow \alpha \in \mathbb{N}] \\ &\wedge \forall\alpha\forall p\forall q[\{P(p) \wedge P(q) \wedge p \text{ divides } q \\ &\wedge p \text{ divides } h - \alpha \wedge q \text{ divides } g \wedge q \text{ divides } h - \alpha\} \\ &\Longrightarrow p \text{ divides } q].\end{aligned}$$

Now we construct an algebra language formula $\mathrm{Seq}(\alpha, n; f, g, h)$ which expresses that α is the nth term of the sequence of constants coded by the triple (f, g, h):

$$\mathrm{Seq}(\alpha, n; f, g, h) \Longleftrightarrow n \in \mathbb{N} \wedge \mathrm{Basis}(f, g, h)$$
$$\wedge \, \alpha \in \mathbb{C} \wedge \exists p[P(p) \wedge p \, \mathrm{divides} \, g \wedge p \, \mathrm{divides} \, h - n$$
$$\wedge \, p \, \mathrm{divides} \, f - \alpha].$$

Using the defining formulas described in Theorem 24.1, we now can say, using a formula in the algebra language, when the sequence coded by (f, g, h) is a Cauchy sequence of rational numbers and when this sequence of rationals converges to 0. (Note that since we have the absolute value function only on $\mathbb{Q}$ at this point, there is no hope of discussing converging or Cauchy sequences outside $\mathbb{Q}$. This is precisely our difficulty in this entire discussion.) For the first of these:

$$\mathrm{Cauchy \; Rat \; Seq}(f, g, h) \Longleftrightarrow \mathrm{Basis}(g, h)$$
$$\wedge \, \forall \alpha \forall n (\mathrm{Seq}(\alpha, n, f, g, h) \Longrightarrow \alpha \in \mathbb{Q})$$
$$\wedge \, \forall \delta \in \mathbb{Q}^{+} \exists m \in \mathbb{N} \forall i, j \in \mathbb{N} \forall \alpha, \beta \in \mathbb{Q}[\{m \leqq i$$
$$\wedge \, m \leqq j \wedge \mathrm{Seq}(\alpha, i, f, g, h) \wedge \mathrm{Seq}(\beta, j, f, g, h)\}$$
$$\Longrightarrow |\alpha - \beta| \leqq \delta].$$

Here we simply have written down in formal terms the usual version of the concept to be defined. ($\mathbb{Q}^{+}$ is the set $\{q \in \mathbb{Q} \mid 0 < q\}$, so it is defined by an algebra language formula.)

For rational sequences converging to 0, we have an entirely similar formula:

$$\mathrm{Zero \; Rat \; Seq} \, (f, g, h) \Longleftrightarrow \mathrm{Basis}(g, h)$$
$$\wedge \, \forall \alpha \forall n (\mathrm{Seq}(\alpha, n, f, g, h) \Longrightarrow \alpha \in \mathbb{Q})$$
$$\wedge \, \forall \delta \in \mathbb{Q}^{+} \exists m \in \mathbb{N} \forall i \in \mathbb{N} \forall \alpha \in \mathbb{Q}[\{m \leqq i \wedge \mathrm{Seq}(\alpha, i, f, gh)\}$$
$$\Longrightarrow |\alpha| \leqq \delta].$$

Consider a pair (g, h) for which Basis (g, h) holds and suppose that f_1, f_2 are functions so that (f_1, g, h) and (f_2, g, h) code Cauchy sequences of rational numbers, say (α_n) and (β_n), respectively. Then it is clear that $(f_1 - f_2, g, h)$ codes the sequence $(\alpha_n - \beta_n)$. [Here it is essential that the same (g, h) be used.] Therefore, (α_n) and (β_n) are equivalent if Zero Rat Seq$(f_1 - f_2, g, h)$ holds.

Next we face the second major difficulty in formalizing our charactrization of $\mathbb{R}$ in the algebra language: We have no direct means of expressing that "the value of f at α equals β," where $f \in \mathcal{E}$ and α, $\beta \in \mathbb{C}$. We approach this indirectly by introducing, as a parameter, a $1 - 1$ conformal mapping σ on $\mathbb{C}$. Of course σ is an element of $\mathcal{E}$ and, as noted above, the property of being a $1-1$ conformal mapping is expressible by an algebra language formula $A(\sigma)$. Now we define a formula in the algebra language, which we will abbreviate by writing $f(\alpha) \underset{\sigma}{=} \beta$, by

$$f(\alpha) \underset{\sigma}{=} \beta \Longleftrightarrow \sigma \text{ is a } 1-1 \text{ conformal map}$$
$$\wedge \alpha \in \text{range}(\alpha) \wedge f(\sigma^{-1}(\alpha)) = \beta$$
$$\Longleftrightarrow A(\sigma) \wedge \sigma - \alpha \text{ is not a unit}$$
$$\wedge \alpha, \beta \in \mathbb{C} \wedge \sigma - \alpha \text{ divides } f - \beta.$$

This formula, which will be quite important in later sections as well, here enables us to express the condition that the composition $f \circ \sigma^{-1}$ carries one sequence coded by (f_1, g, h) to a second sequence coded by (f_2, g, h). The formula that expresses this is the following:

$$\forall n \in \mathbb{N} \forall \alpha, \beta \in \mathbb{C}[\text{Seq}(\alpha, n, f_1, g, h) \wedge \text{Seq}(\beta, n, f_2, g, h) \Longrightarrow f(a) \underset{\sigma}{=} \beta].$$

Note that for a fixed $1-1$ conformal mapping σ, as f ranges over $\mathcal{E}$ the composition mapping $f \circ \sigma^{-1}$ ranges over $\mathcal{E}$.

We now are ready to give the formula in the algebra language which defines $\mathbb{R}$ in $\mathcal{E}$. (It is convenient to define $\mathbb{C}\backslash\mathbb{R}$ instead.) We see that $\alpha \in \mathbb{C}\backslash\mathbb{R} \Longleftrightarrow \alpha \in \mathbb{C}$ and there is a $1-1$ conformal mapping σ on $\mathbb{C}$, and for any Cauchy sequence (α_n) of rationals that is $f \in \mathcal{E}$ and a Cauchy Sequence (β_n) of rationals with the properties

(i) $f(\alpha_n) \underset{\sigma}{=} \beta_n$ for all $n \in \mathbb{N}$);

(ii) (α_n) and (β_n) are equivalent Cauchy sequences of rationals;

(iii) $f(\alpha) \underset{\sigma}{=} \alpha$ is false.

It remains only to show that this equivalence is correct. Earlier we proved the equivalence $\alpha \in \mathbb{C}\backslash\mathbb{R} \Longleftrightarrow \alpha \in \mathbb{C}$, and for any Cauchy sequence (α_n) of rationals there is an entire function g and a Cauchy sequence (β_n) of rationals with the properties

(i') $g(\alpha_n) = \beta_n$ for all n;

(ii') (α_n) and (β_n) are equivalent Cauchy sequences of rationals;

(iii') $g(\alpha) \neq \alpha$.

Clearly, if σ exists for α as above, and if (α_n), (β_n) and f satisfy (i), (ii), and (iii), then we need only set g equal to $f \circ \sigma^{-1}$. Conversely, the range of every $1-1$ conformal mapping σ on $\mathbb{C}$ includes α and $\mathbb{Q}$. Given such a σ and g satisfying (i'), (ii'), and (iii'), just take f to be the composition $g \circ \sigma$.

This finally completes the proof that $\mathbb{R}$ is definable in $\mathcal{E}$. As was discussed earlier, from this we get formulas defining $\leq$ on $\mathbb{R}$ and the absolute value on $\mathbb{C}$. Thus the proof of Theorem 23.2 is complete.

Theorem 23.2 is fundamental to nearly all of our other results.

More on the Algebra Language

In this section we present a variety of results which, when taken together, show that the expressive power of the algebra language is strong enough to include all of the classical mathematical theory of entire functions. It is quite striking that this should be possible, since the first-order algebra language seems to be so limited. These results are not used in any other part of this chapter, and we have not made a great effort to be complete nor to give more than sketchy proofs. Our goal is to give examples that show what is possible.

We discussed how to deal with sequences of constants within the first-order theory of $\mathcal{E}$. Now we will improve this to code sequences of *functions*. The idea is to fix a point z_0 in $\mathbb{C}$ and a sequence $\{z_n\}$ in $\mathbb{C}$ that converges to z_0. By the uniqueness of analytic continuation, a function $f \in \mathcal{E}$ is uniquely determined by the sequence of constants $\{f(z_n)\}$. Hence, a sequence $\{f_m\}$ of functions can be coded by an infinite matrix of constants $\{f_m(z_n) \ m, n \in \mathbb{N}\}$. Now this is not quite enough, since we cannot refer to the points of $\mathbb{C}$ in a direct way. We overcame this earlier by introducing a parameter σ that is a $1-1$ conformal mapping on $\mathbb{C}$. We replace the points z_n by the constants $\alpha_n = \sigma(z_n)$ and refer to the values $\beta_{m,n} = f_m(z_n)$ by using the equivalent, definable relationship $f_m(\alpha_n) \underset{\sigma}{=} \beta_{m,n}$. Finally, we code the sequence (α_n) and the matrix (β_{mn}) into a single matrix that has $(\alpha - n)$ as the top row.

To code an infinite matrix of constants we proceed as earlier, but use a basis triple (g, h_1, h_2) instead of a pair (g, h). Here we require that $g(\neq 0)$ have an infinite zero set in $\mathbb{C}$, as before, and that (h_1, h_2) map this zero set bijectively onto $\mathbb{N} \times \mathbb{N}$. It is routine to construct a formula $M \operatorname{Basis}(x, y, z)$ in the algebra language such that $M \operatorname{Basis}(g, h_1, h_2)$ holds in $\mathcal{E}$ if and only if this basis condition is satisfied.

When $M \operatorname{Basis}(g, h_1, h_2)$ is true, any function f in $\mathcal{E}$ determines a matrix of constants (α_{mn}) by taking α_{mn} to be the value of f at the unique zero z_{mn} of g for which $h_1(z_{mn}) = m$ and $h_2(z_{mn}) = n$. Earlier we expressed this relation using point functions. Also, by the interpolation theorem for entire functions, every matrix (α_{mn}) is coded in this way by some f, no matter which basis triple (g, h_1, h_2) is used.

The matrices (α_{mn}) that arise in coding sequences of functions are not arbitrary, of course. First, the sequence (α_{mn}) from the top row must be a Cauchy sequence in $\mathbb{C}$. Finally, for each $m > 0$ there must exist a function f_m in $\mathcal{E}$ such that

$$f_m(\alpha_{0n}) \underset{\sigma}{=} \alpha_{mn} \quad \text{for all} \quad n \in \mathbb{N}.$$

Evidently, f_m is uniquely determined by this condition. It is the nth function in the sequence coded by (α_{mn}) and σ. As was discussed previously, we take the matrix (α_{mn}) to be coded by some (f, g, h_1, h_2), where $M \operatorname{Basis}(g, h_1, h_2)$, holds.

We can construct an algebra formula $\text{Code}(f, g, h_1, h_2, \sigma)$ that holds in $\mathcal{E}$ if and only if σ is a $1-1$ conformal mapping on G and $M\,\text{Basis}(g, h_1, h_2)$ holds and the matrix (α_{mn}) coded by f using the basis (g, h_1, h_2) satisfies the conditions above. [That is, the sequence α_{0n} and its limit lie in the range of σ and, for each $m > 0$, there is an f_m in $\mathcal{E}$ so that $f_m(\alpha_{0m}) \underset{\sigma}{=} \alpha_{mn}$ for all $n \in \mathbb{N}$.] Then we formulate an algebra formula $F\,\text{Seq}(k, m, f, g, h_1, h_2, \sigma)$ that holds in $\mathcal{E}$ if and only if $\text{Code}\,(f, g, h_1, h_2, \sigma)$ holds, $m \in \mathbb{N}$, and k is the (unique) function that satisfies $k(\alpha_{0n}) \underset{\sigma}{=} \alpha_{mn}$ for all $n \in \mathbb{N}$. That is, $F\,\text{Seq}(k, m, f, g, h_1, h_2, \sigma)$ holds if and only if k is the mth function in the sequence coded by (f, g, h_1, h_2, σ).

Using the absolute value on complex constants and the "evaluation" of functions f via σ (that is, in the form $f(\alpha) \underset{\sigma}{=} \beta$), we now may obtain algebra formulas that express various types of convergence of the coded sequence of functions. This can be done for pointwise convergence, uniform convergence on $\mathbb{C}$, or even uniform convergence on compact subsets of $\mathbb{C}$. For this last type of convergence it is not necessary to quantify over arbitrary compact sets, but rather to use a particular exhaustion of $\mathbb{C}$ by compact sets. For example, suppose $\text{Code}\,(f, g, h_1, h_2, \sigma)$ holds in $\mathcal{E}$ and we wish to discuss convergence of the coded sequence of functions (f_m). Consider the sets $G'_n \subseteq \text{range}(\sigma)$ defined by

$$G'_n = \{\alpha \in \mathbb{C} \mid |\alpha| \leqq n\}.$$

These sets are definable using an algebra formula from the parameters n and σ. Also, (f_m) converges uniformly on compact subsets of $\mathbb{C}$ if and only if $(f_m \circ \sigma^{-1})$ converges uniformly on each of the compact sets G'_n. This can be expressed by an algebra formula using $F\,\text{Seq}$ at the end to replace mention of (f_m).

This method of representing sequences of functions within $\mathcal{E}$ enables us to define many specific sequences—for example, the sequence of powers (f^m) of a particular function. From this we can find an algebra formula which expresses that g is equal to a polynomial in f.

Next we discuss a method for interpreting in $\mathcal{E}$ the lattice of open subsets of $\mathbb{C}$. (This procedure is also used later, where it is discussed in greater detail.) This is done by associating each open set $\mathcal{O}$ with the set $\mathcal{R}(\mathcal{O})$ of all pairs (q, r), where q is a Gaussian rational, r is a rational > 0, and the disc $\{\alpha \in \mathbb{C} \mid |\alpha - q| < r\}$ is contained in $\mathcal{O}$. Since $\mathcal{O}$ is the union of this family of discs, we see that $\mathcal{O} \neq \mathfrak{O}$ implies $\mathcal{R}(\mathcal{O}) \neq \mathcal{R}(\mathfrak{O})$ for open sets $\mathcal{O}$, $\mathfrak{O}$. The set $\mathcal{R}(\mathcal{O})$ can be coded by a quadruple (f_1, f_2, g, h) by first writing $\mathcal{R}(\mathcal{O})$ as a sequence (q_n, r_n) for $n \in \mathbb{N}$ and then taking (q_n, r_n) to be the value of $(f_1(z), f_2(z))$ at the unique $z \in \mathbb{C}$ for which $g(z) = 0$ and $h(z) = n$. Here, (g, h) satisfies the Basis formula described above.

We first obtain an algebra formula $\mathcal{O}\,\text{Basis}(f_1, f_2, g, h)$ that is true in $\mathcal{E}$ if and only if there is an open set $\mathcal{O}$ such that $\mathcal{R}(\mathcal{O}) = \{\alpha_n, \beta_n) \mid n \in \mathbb{N}\}$, where (α_n, β_n) is the sequence of pairs coded by (f_1, f_2) using the basis pair (g, h). This must express that every α_n is in $\mathbb{Q}(i)$, every β_n is in $\mathbb{Q}^+$; also, it must express that if $q \in \mathbb{Q}(i)$, $r \in \mathbb{Q}^+$ and if the disc $\{\gamma \in \mathbb{C} \mid |\gamma - q| < r\}$ is contained in the union of the discs $\{\gamma \in \mathbb{C} \mid |\gamma - \alpha_n| < \beta_n\}$, then (q, r) is in the set $\{(\alpha_n, \beta_n) \; n \in \mathbb{N}\}$. Of course, $\mathcal{O}$ is just the union of the discs $\{\gamma \in \mathbb{C} \mid |\gamma - \alpha_n| < \beta_n\}$ for $n \in \mathbb{N}$. Thus we can obtain an algebra formula $\mathrm{Open}(\alpha, f_1, f_2, g, h)$ which expresses that $\mathcal{O} \, \mathrm{Basis}(f_1, f_2, g, h)$ holds and that α is in the open set $\mathcal{O}$ for which $\mathcal{R}(\mathcal{O}) = \{(\alpha_n, \beta_n) \mid n \in \mathbb{N}\}$.

Evidently we also can represent sequences of open sets by using matrix pairs $(\alpha_{mn}, \beta_{mn})$ of constants, each row of which codes an open subset of $\mathbb{C}$. This gives us another way of referring to an exhaustion (G'_n) of $\mathbb{C}$ by compact sets, referring instead to the sequence $\mathbb{C}\backslash G'_n)$ of open sets. Also, we can develop a way of representing all analytic subsets of $\mathbb{C}$ (including all Borel sets) using the Souslin operation applied to sequences of sets. In addition, we can construct an algebra formula that expresses the value of Lebesgue measure applied to these analytic sets using approximation by open and closed sets. These matters are more complicated, and we omit the details.

Finally, we discuss how to express integration of functions and the "Nevanlinna characteristic" applied to entire functions using algebra formulas. First we consider integration. Fix $f \in \mathcal{E}$ and let σ be a $1 - 1$ conformal mapping on $\mathbb{C}$. We then can express by using algebra formulas how to integrate $f \circ \sigma^{-1}$ along a circular curve $\gamma = \{z \mid |z - \alpha| = r\}$ contained in the range of σ. That is, we have an algebra formula $\mathrm{Int}(f, \sigma, \alpha, r, \beta)$ which holds in $\mathcal{E}$ if and only if σ, α, r are as above and $\int_\gamma f dz = \beta$. This can be done by considering a sequence of successively finer subdivisions of γ and by evaluation of $f \circ \sigma^{-1}$ on points of γ using our expressions $f(\eta) = \delta \atop \sigma$ as above. We then get upper and lower sums and evaluate the integral by taking limits. (Lebesgue integration can be handled also, but it is more complicated. In any case, our functions are highly continuous.) By a similar procedure we can also treat integration over the *inside* of the curve γ, with respect to area measure.

Now we consider the "Nevanlinna characteristic" on entire functions. For an entire function f this is defined by

$$T(r, f) = \frac{1}{2\pi} \int_0^{2\pi} \log^+ |f(re^{i\theta}| d\theta.$$

We also consider the order α of f defined by

$$\alpha = \limsup_{r \to \infty} \frac{\log T(r, f)}{\log r}.$$

Note that when we apply the above approach to integration, the $1 - 1$ conformal mapping σ that appears as a parameter must be affine. Also, the order of f is equal to the order of $f \circ \sigma^{-1}$, no matter which affine σ we choose. This shows that there is an algebra formula $\mathrm{Ord}(f, \alpha)$ which is true in the algebra of entire functions if and only if the order of f equals α.

Obviously there is no completely natural point at which to stop this general discussion of how to express the mathematics of entire functions in the first-order theory of $\mathcal{E}$. As far as we can see, essentially all of the classical theory of entire functions, including topological and measure-theoretic aspects, can be so represented.

Derivatives and Definable Constants

We will call a constant $\alpha \in \mathbb{C}$ *definable* if there is a formula $D(x)$ in the algebra language such that, for every function $f \in \mathcal{E}$,

$$D(f) \text{ holds in } \mathcal{E} \Longleftrightarrow f \text{ equals the constant } \alpha.$$

Evidently, i is a definable constant, since we have included a name for it in the algebra language. Also, it is clear that α is definable if and only if both the real part of α and the imaginary part of α are definable.

One way to obtain a large number of definable constants is via infinite series and the device for coding sequences of constants that was discussed earlier. Recall that such a sequence is coded by a triple of functions (f, g, h): α is the nth term of the sequence if $f(z) = \alpha$, where z is the (unique) zero of g for which $h(z) = n$. It is now a routine matter, given the definability results obtained earlier, to construct an algebra language formula $\mathrm{Series}(x, y, z, w)$ such that, for any $\alpha, f, g, h \in \mathcal{E}$,

$$\mathrm{Series}(\alpha, f, g, h) \text{ holds in } \mathcal{E} \text{ if and only if } (f, g, h) \text{ codes}$$

$$\text{a sequence } (\alpha_n) \text{ and } \sum \alpha_n \text{ converges to } \alpha.$$

To make this clearer, we take the first step toward the formula:

$$\mathrm{Series}(\alpha, f, g, h) \Longleftrightarrow \mathrm{Basis}(g, h) \wedge \alpha \in \mathbb{C}$$
$$\wedge \, (\exists k)[\forall n \in \mathbb{N} \forall \beta, \gamma, \delta \in \mathbb{C}(\mathrm{Seq}(\beta, n, k, g, h)$$
$$\wedge \, \mathrm{Seq}(\gamma, n + 1, f, g, h) \Longrightarrow \mathrm{Seq}(\beta + \gamma, n + 1, k, g, h))$$
$$\wedge \, \forall \beta \in \mathbb{C}(\mathrm{Seq}(\beta, 0, f, g, h) \Longrightarrow \mathrm{Seq}(\beta, 0, k, g, h))$$
$$\wedge \, \forall \delta \in \mathbb{R}^+ \exists n \in \mathbb{N} \, \forall m \in \mathbb{N} \, \forall \beta \in \mathbb{C}(n \leqq m \wedge \mathrm{Seq}(\beta, m, k, g, h)$$
$$\Longrightarrow |\beta - \alpha| \leqq \delta)].$$

The middle two clauses of this formula assert that the sequence (β_n) coded by the triple (k, g, h) is the sequence of partial sums of the sequence (α_n) coded by (f, g, h). The third clause asserts that (β_n) converges to α.

Now we will show, as an example, that e is a definable constant. This follows because we can say in a first-order way that (f, g, h) codes a sequence

(α_n) which satisfies the recurrence relation $\alpha_0 = 1$, $\alpha_{n+1} = \alpha_n/(n+1)$, forcing $\alpha_n = 1/n!$ and so $e = \sum \alpha_n$. Thus the defining formula $D(x)$ for e is obtained from:

$$D(\alpha) \Longleftrightarrow \exists f \exists g \exists h [\text{Series}(\alpha, f, g, h) \wedge \text{Seq}(1, 0, f, g, h)$$
$$\wedge \forall n \in \mathbb{N} \ \forall \beta \in \mathbb{C}(\text{Seq}(\beta, n, f, g, h)) \Longrightarrow \text{Seq}(\beta/n + 1., n + 1, f, g, h)].$$

[Strictly speaking, we cannot use division, but this is easily eliminated from $D(x)$.]

Clearly this same device can be used to obtain most of the familiar transcendental real numbers, as well as many others, for example, Liouville numbers such as $\sum 10^{n!}$. All that is required is that the number be the limit of a series whose terms are generated using some recurrence formula.

Theorem 23.3. *Let F be the set of all definable constants from $\mathbb{C}$. Then F is a countable algebraically closed field that contains e and π.*

Proof. It is clear that F is a subfield of $\mathbb{C}$. For example, if $D(x)$ defines $\alpha = 0$, then the formula

$$\exists y (D(y) \wedge xy = 1)$$

defines $1/\alpha$. Also, F is countable, since there are only countably many formulas in the algebra language. It is easy to show that F is algebraically closed. For example, suppose α_j is definable by $D_j(x)$ for $j = 0, 1, 2, 3$. We will show how to define a root of the polynomial $p(z) = \alpha_0 z^3 + \alpha_1 z^2 + \alpha_2 z + \alpha_3$. Consider the linear ordering on $\mathbb{C}$ defined by taking

$$\alpha \leqq \beta \Longleftrightarrow \Re(\alpha) \leqq \Re(\beta) \quad \text{or}$$
$$(\Re(\alpha) = \Re(\beta) \quad \text{and} \quad \Im(\alpha) \leq \Im(\beta)).$$

This ordering is definable by an algebra formula $B(x, y)$ in the sense that for any $\alpha, \beta \in \mathbb{C}$

$$B(\alpha, \beta) \quad \text{holds in} \quad \mathcal{E} \Longleftrightarrow \alpha \leqq \beta.$$

Using this we can define the "smallest" root of $p(z)$ by a formula $D(x)$. Namely, $D(\alpha)$ is equivalent to

$$\exists \alpha_0 \exists \alpha_1 \exists \alpha_2 \exists \alpha_3 [D_0(\alpha_0) \wedge D_1(\alpha_1) \wedge D_2(\alpha_2)$$
$$\wedge D_3(\alpha_3) \wedge \alpha \in \mathbb{C} \wedge \alpha_0 \alpha^3 + \alpha_1 \alpha^2 + \alpha_2 \alpha + \alpha_3 = 0$$
$$\wedge \forall \beta (\beta \in \mathbb{R} \wedge \alpha_0 \beta^3 + \alpha_1 \beta^2 + \alpha_2 \beta + \alpha_3) = 0 \Longrightarrow B(\alpha, \beta)].$$

Other interesting definability results for the constants come from an indirect treatment of derivatives, which we now discuss. Of course, there is no hope of obtaining a first-order definition of the relation between a function and its derivative, since this relation is not conformally invariant. Our indirect approach comes via the use of a parameter σ, which is a $1-1$ conformal mapping, as was used earlier to obtain the definition of $\mathbb{R}$. Given such a σ and given f, $g \in \mathcal{E}$, we will show that the relation $(g \circ \sigma^{-1}) = (f \circ \sigma^{-1})'$ is a first-order property of the triple (f, g, σ). Namely, this relation is equivalent to the condition:

$$\forall \alpha \in \mathbb{C} \; \exists h \exists k \exists \beta \in \mathbb{C} \; \exists \gamma \in \mathbb{C}q \; [\text{if } \sigma - \alpha \text{ is a nonunit, then}$$
$$g = \beta + (\sigma - \alpha)h \quad \text{and} \quad f = \gamma + \beta(\sigma - \alpha) + (\sigma - \alpha)^2 k].$$

Proof. ($\Longleftarrow$) Let $\alpha = \sigma(z)$ for some $z \in \mathbb{C}$ and suppose the equations $g = \beta + (\sigma - \alpha)h$ and $f = \gamma + \beta(\sigma - \alpha) + (\sigma - \alpha)^2 k$ hold over $\mathbb{C}$. Substitute σ^{-1} in the second equation and get

$$f \circ \sigma^{-1}(w) = \gamma + \beta(w - \alpha) + (w - \alpha)^2 k(\sigma^{-1}(w)) \quad \text{for all} \quad w \in \mathbb{C}.$$

Differentiating with respect to w and setting $w = \alpha$ yield $(f \circ \sigma^{-1})'(\alpha) = \beta$. Since $(g \circ \sigma^{-1})(\alpha) = \beta$, we have the desired equation.

($\Longrightarrow$) Suppose that $(g \circ \sigma^{-1}) = (f \circ \sigma^{-1})'$ and consider $\alpha \in \mathbb{C}$. Expanding $g \circ \sigma^{-1}$ and $f \circ \sigma^{-1}$ about the point α yields

$$(g \circ \sigma^{-1})(w) = \beta + (w - \alpha)h(w),$$
$$(f \circ \sigma^{-1})(w) = \gamma + \beta(w - \alpha) + (w - \alpha)^2 k(w).$$

Substituting $w = \sigma(z)$ yields the equations needed.

The exponential function is uniquely determined on any neighborhood of 0 by its functional equation $f' = f$ and $f(0) = 1$. This leads to a certain definability result for the exponential function:

Theorem 23.4. *There is a formula $E(x, y)$ in the algebra language such that for any α, $\beta \in \mathcal{E}$*

$$E(\alpha, \beta) \text{ holds in } \mathcal{E} \text{ if and only if } \alpha, \; \beta \in \mathbb{C} \text{ and } e^\alpha = \beta.$$

Proof. Given α, $\beta \in \mathbb{C}$, consider the following condition:

$E(\alpha, \beta) \Longleftrightarrow$ there exists a $1 - 1$ conformal mapping σ and a function f, both in $\mathcal{E}$, such that
   (i) $f(0) \underset{\sigma}{=} 1$,
   (ii) $(f \circ \sigma^{-1})' = (f \circ \sigma^{-1})$,
   (iii) $f(\alpha) \underset{\sigma}{=} \beta$.

From our previous discussion it is clear that this can be expressed by an algebra language formula. Parts (i) and (ii) imply that $f \circ \sigma^{-1}$ is equal to the exponential function, and part (iii) therefore says that $e^\alpha = \beta$. Thus, $E(\alpha, \beta)$ implies $e^\alpha = \beta$.

Conversely, given that $e^\alpha = \beta$, to satisfy $E(\alpha, \beta)$ we need only set σ equal to any $1 - 1$ conformal mapping on $\mathbb{C}$ and then set $f(z) = e^{\sigma(z)}$ for $z \in \mathbb{C}$.

Corollary 23.5. *Let F_0 be the field of definable constants. Then F_0 is closed under the exponential function.*

Note that Corollary 23.5 gives an alternate proof that F_0 contains e. Also, it yields that F_0 has infinite transcendence degree. Indeed, by the Hermite-Lindemann Theorem, if $\alpha_1, \ldots, \alpha_n$ are algebraic numbers (hence in F_0) that are linearly independent over $\mathbb{Q}$, then $e^{\alpha_1}, \ldots, e^{\alpha_n}$ are algebraically independent elements of F_0. (See [21].)

We close this section by noting that the other elementary functions, such as $\sin(z)$ and $\cos(z)$, can be treated in a way that is similar to our discussion of e^z.

Recursive Undecidability

Earlier there was presented a way of defining $\mathbb{N}$ in $\mathcal{E}$ and a method for coding countable sets of constants. This yields an effective interpretation of second-order number theory in the first-order theory of $\mathcal{E}$.

This interpretation is an example of a $1-1$ *reduction* of one "problem" or set to another. If $\mathfrak{S}_1$, $\mathfrak{S}_2$ are sets of sentences in formal languages L_1, L_2, respectively, we say $\mathfrak{S}_1$ is $1-1$ reducible to $\mathfrak{S}_2$, and write $\mathfrak{S}_1 \leqq_1 \mathfrak{S}_2$, if there is an effectively computable $1-1$ function φ (i.e., a recursive function) such that for any sentence S of L_1,

$$S \in \mathfrak{S}_1 \Longleftrightarrow \varphi(S) \in \mathfrak{S}_2.$$

We say that $\mathfrak{S}_1$ and $\mathfrak{S}_2$ are *recursively isomorphic* if there is an effectively computable function φ that maps the sentences of L_1 bijectively onto the sentences of L_2 and which satisfies $\varphi(\mathfrak{S}_1) = \mathfrak{S}_2$. Evidently this means that the problems of deciding membership in $\mathfrak{S}_1$ and in $\mathfrak{S}_2$ are effectively equivalent in a strong way. It is a well-known fact [34] due to Myhill that if $\mathfrak{S}_1 \leqq_1 \mathfrak{S}_2$ and $\mathfrak{S}_2 \leqq_1 \mathfrak{S}_1$, then $\mathfrak{S}_1$, $\mathfrak{S}_2$ are recursively isomorphic. (The converse is obvious.)

As an example of what type of undecidability theorem can be proved using the results above, we consider the ring of entire functions $\mathcal{E}$. Robinson [33] showed that the first-order theory of this ring is undecidable by showing that first-order number theory can be interpreted in it. The following result is a substantial improvement of this.

Theorem 23.6. *The first-order theory of the ring of entire functions is recursively isomorphic to second-order number theory.*

Proof. Let $\mathfrak{S}_1$ denote the first-order theory of $\mathcal{E}$ and let $\mathfrak{S}_2$ denote the second-order theory of $(\mathbb{N}, +, \cdot)$. We will show $\mathfrak{S}_1 \leqq_1 \mathfrak{S}_2$ and $\mathfrak{S}_2 \leqq_1 \mathfrak{S}_1$. Let $\mathfrak{S}_1'$ denote the first-order theory of the algebra of entire functions $\mathcal{E}$. Earlier it was shown that the field of constants is definable in $\mathcal{E}$, which yields a direct interpretation of $\mathfrak{S}_1'$ into $\mathfrak{S}_1$. In particular, $\mathfrak{S}_1' \leqq_1 \mathfrak{S}_1$. As was discussed earlier in this section, we also have a direct interpretation of second-order number theory in $\mathfrak{S}_1'$. This implies $\mathfrak{S}_2 \leqq_1 \mathfrak{S}_1'$, so that $\mathfrak{S}_2 \leqq_1 \mathfrak{S}_1$ has been proved.

To show $\mathfrak{S}_1 \leq_1 \mathfrak{S}_2$, we sketch how to give an interpretation of the first-order theory of $\mathcal{E}$ in second-order number theory. Each entire function f has a power series representation centered at 0,

$$f(z) = \sum \alpha_n z^n,$$

with an infinite radius of convergence. Each coefficient $a_n = x_n + y_n i$ can be identified with two sequences of integers, say $\{r_{n,j} \mid j \in \mathbb{N}\}$ and $\{s_{n,j} \mid j \in \mathbb{N}\}$, where $r_{n,0}$ is the integer part of x_n and $r_{n,j}$ is the jth decimal digit of x_n, and similarly for $s_{n,j}$ and y_n. Therefore, f can ultimately be identified with an infinite matrix of integers $M = (m_{ij} \mid i, j \in \mathbb{N})$ obtained by setting $m_{i,2j} = r_{ij}$ and $m_{i,2j+1} = s_{ij}$ for all $i, j \in \mathbb{N}$. Finally, M can be converted to a set of integers $M^\# = \{2^i \cdot 3^j \cdot 5^{m_{ij}} \mid i, j \in \mathbb{N}\}$. Evidently we can recover the function f from the set $M^\#$. To show that this provides the desired interpretation, one should prove that the collection of all sets of the form $M^\#$ is definable in the second-order theory of $(\mathbb{N}, +, \cdot)$ and prove the same for the relations on these sets which correspond to addition and multiplication of entire functions. The details are tedious and routine, and we choose to omit them.

Corollary 23.7. *Second-order number theory is $1 - 1$ reducible to T_0.*

24
Identities of Exponential Functions

In this chapter, which is based on [13], we take up some questions prompted by mathematical logic, notably Tarski's "High School Algebra Problem." We study identities between certain functions of many variables that are constructed by using the elementary functions of addition $x + y$, multiplication $x \cdot y$, and one-place exponentiation e^x, starting out with all the complex constants and the independent variables $z_1, \ldots, z_n$. We show that every true identity in this class follows from the natural set of 11 axioms of High School Algebra. The major tool in our proofs is the Nevanlinna theory of entire functions of n complex variables, of which we give a brief sketch. It is entirely parallel to the one-variable theory presented in detail earlier in this book. The timid reader can take $n = 1$, at least for a first reading.

Tarski's conjecture for a more extended class of terms than those we consider here has been shown to be false by Wilkie (see [50]). The largest class that we are aware of for which Tarski's axioms have been shown to be complete was studied in [11].

We briefly recapitulate the basic Nevanlinna theory.

Consider first the case of one variable, $n = 1$. For meromorphic functions f of one variable defined on the complex plane $\mathbb{C}$, the characteristic function is defined for $0 < r < \infty$ by

$$T(r, f) = m(r, f) + N(r, f);$$

this is a sum of two terms: the proximity function

$$m(r, f) = \frac{1}{2\pi} \int_{-\pi}^{\pi} \log^+ |f(re^{i\theta})| d\theta,$$

which measures how close, on the average, f is to ∞, and the average counting function

$$N(r, f) = \int_0^r \frac{n(t, f)}{t}\, dt,$$

where $n(r)$ is the number of poles of f in the disc $|z| \leq r$. Here, r ranges over the interval $0 < r < \infty$, and appropriate modifications must be made in the definitions of $N(r, f)$ if $f(0) = 0$ or if $f(0) = \infty$. The function $\log^+(t)$ is defined by setting $\log^+(t) = \log(t)$ for $t \geq 1$ and $\log^+(t) = 0$ for $0 \leq t \leq 1$. The growth of the characteristic $T(r, f)$ as $r \to \infty$ gives a very useful measure of the growth of f. The basic properties that we shall use are listed below.

(C2.0) $T(r, f)$ is a nondecreasing function of r and a convex function of $\log r$.

(C2.1) $T(r, f + g) \leq T(r, f) + T(r, g) + O(1)$.

(C2.2) $T(r, fg) \leq T(r, f) + T(r, g)$.

(C2.3) $T(r, 1/(f - a)) = T(r, f) + O(1)$ for any complex constant a.

(C2.4) $T(r, f/g) \leq T(r, f) + T(r, g) + O(1)$.

(C2.5) $T(r, e^g)/T(r, g) \to \infty$ as $r \to \infty$ if $T(r, g)$ is unbounded.

The other basic fact we need about the characteristic $T(r, f)$ is the Lemma of the Logarithmic Derivative (LLD):

(C2.6)

$$m(f, f'/f) \leq O\left(\log(T(r, f))\right) + \log r\right),$$

except possibly for r lying in a set E of finite length.

When it comes to several variables, the theory is substantially the same and the basic properties we need are still expressed in the same form (C2.0)–(C2.6). (See [42] for the details of proofs.) Alternatively, one could use a characteristic based on the exhaustion of $\mathbb{C}^n$ by balls, rather than by polydisks $\mathbb{D}^n$. (See [48] and [10].) For a meromorphic function $f(z_1, \ldots, z_n)$ defined on $\mathbb{C}^n$, we define

$$m(r, f) = \frac{1}{(2\pi)^n} \int_{-\pi}^{\pi} \cdots \int_{-\pi}^{\pi} \log^+ |f(re^{i\phi_1}, \ldots re^{i\phi_n})| d\phi_1 \cdots d\phi_n,$$

where $0 < r < \infty$. Also, $N(r, f)$ is defined much as before, as an averaged counting function of the poles of f. Note that if f is a holomorphic function on $\mathbb{C}^n$, then $T(r, f) = m(r, f)$.

(In [42], a characteristic $T(\vec{r}, f)$ is developed for a vector variable $\vec{r} = (r_1, \ldots, r_n)$, but we use only the diagonal case $r_1 = \cdots = r_n = r$.)

The basic properties above, including LLD, are shown to hold in [42] or follow exactly as in the case of one variable (e.g., (C2.5)). One thing which needs explanation is the derivative f' that occurs in the LLD. When $n \geq 2$, we shall take f' to stand for the Euler operator

$$f' = Df = z_1 \frac{\partial f}{\partial z_1} + \cdots + z_n \frac{\partial f}{\partial z_n}.$$

This has the useful property that $Df = 0$ if and only if f is identically constant. (This is because f may be expanded as a nicely converging sum of homogeneous polynomials and because $DP = mP$ for any homogeneous polynomial of degree m.)

Our basic tool is a lemma proved in one dimension by Hiromi and Ozawa—see Chapter 17 of this book. The proof in n dimensions is similar to the proof in 1 dimension and depends only on the properties (C2.0)-(C2.6) we have listed of the characteristic function $T(r, f)$. (Carlos Berenstein has recently pointed out that the proof in [13] is incomplete. The authors of [13] are preparing a complete proof. More varied Wronskians are needed to establish linear dependence in the N-dimensional version of the Hiromi-Ozawa lemmal. See [4] for a correct version of the proof for the ball-characteristic.)

Lemma 24.1. (Hiromi-Ozawa). *Let* $a_0(z), \ldots, a_n(z)$ *be meromorphic functions and let* $g_1(z), \ldots, g_n(z)$ *be holomorphic functions defined on the domain* $\mathbb{C}^N$. *Suppose that these functions satisfy*

(a)
$$T(r, a_j) = o\left(\sum_{i=1}^{n} m(r, e^{g_1})\right)$$

for each $j = 0, 1, \ldots, n$; *and*

(b)
$$T(r, e_1^g) \neq O(\log r)$$

for at least one $i = 1, 2, \ldots, n$. *Under these hypotheses, if the identity*

$$\sum_{j=1}^{n} a_j(z)e^{g_j(z)} = a_0(z)$$

holds for $z \in \mathbb{C}^N$, *then there exists constants* $c_1, \ldots, c_n$ *(not all 0) so that*

$$\sum_{j=1}^{n} c_j \cdot a_j(z)e^{g_j(z)} = 0$$

for all $z \in \mathbb{C}^N$.

Our use of the Nevanlinna theory tools previously discussed comes entirely through this Hiromi-Ozawa Lemma. Indeed, we use it only in cases where $g_1, \ldots, g_n$ are holomorphic functions, so that $T(r, e^{g_1}) = m(r, e^{g_1})$ (and the prohibition that the function not take the value 0 at the origin is satisfied), and in cases where $a_0, a_1, \ldots, a_n$ are slowly growing functions. Here we consider expressions that are built up from variables and complex constants using addition, multiplication, and the 1-variable exponential function e^x (where e is the usual base of the natural logarithm).

We prove a version of Tarski's High School Algebra Conjecture for these expressions. (This result was proved independently by van den Dries [47] and, for terms containing just one variable, by Wilkie [49]. Their methods are quite different from ours.) We also settle positively a conjecture, due to Schanuel, which asserts that if f is a function on $\mathbb{C}^n$ which is defined by an expression of this type, and if f is nowhere equal to 0, then $f = e^g$ for a function g on $\mathbb{C}^n$ which is also defined by an expression of the kind considered here.

Definition 24.2. Σ is the smallest class of terms which contains the variables x_1, $x_2, \ldots$ and a constant for each complex number, and which contains the terms $s + t$, $s \cdot t$ and $\exp(t)$ for each $s, t \in \Sigma$.

Here we interpret $\exp(t)$ to stand for e^t. We note that if $t \in \Sigma$ and the variables of t are among $x_1, \ldots, x_n$, then t defines a holomorphic function on all of $\mathbb{C}^n$. If $s \in \Sigma$ also has its variables among $x_1, \ldots, x_n$, we write $t \equiv s$ to mean that t and s define the same function on $\mathbb{C}^n$. (Various equivalent formulations of this definition are possible in special cases because of the uniqueness of holomorphic functions. For example, if t and s contain only real constants, we may be interested only in the functions they define on $\mathbb{R}^n$. But $t \equiv s$ will hold as long as t and s define the same function on $\mathbb{R}^n$, or even on S^n where $S \subseteq \mathbb{C}$ is any set with a limit point in $\mathbb{C}$.)

One has the additional useful fact that a holomorphic function f on $\mathbb{C}^n$ has the small characteristic $T(r, f) = O(\log(r))$ if and only if f is a polynomial. Hence, if f is a polynomial and g is any nonconstant holomorphic function on $\mathbb{C}^n$, then $T(r, f) = o(T(r, e^g))$ (which is necessary as part of the hypotheses of the Hiromi-Ozawa Lemma as we apply it). See [17, Proposition 4.4ff]. For holomorphic functions this can be proved by estimating the Poisson integral for $\log|f|$ to show that f is of polynomial growth as a function of x_j (when x_i, $i \neq j$, are held fixed), for each $j = 1$, $2, \ldots, n$. By the Liouville Theorem in one variable, then, f is a polynomial separately in each x_j. That f is globally a polynomial now follows from [30]. (There must be many other proofs of our assertion in the literature.)

Theorem 24.3.   (Tarski's Conjecture for Σ). *If t, s are any two terms in Σ and $t \equiv s$, then the identity $t = s$ is probable from the axioms*

$$x + (y + z) = (x + y) + z, \qquad x(yz) = (xy)z,$$
$$x + y = y + x, \qquad\qquad xy = yx,$$
$$x + 0 = x, \qquad\qquad 1 \cdot x = x,$$
$$x(y + z) = xy + xz, \qquad\qquad 0 \cdot x = 0,$$

$$\exp(x + y) = \exp(x) \cdot \exp(y),$$

together with all axioms giving the facts of addition, multiplication, and exponentiation for constants from $\mathbb{C}$.

Proof. Because we have included here a constant for -1, the operation of subtraction is available and we need only consider the case where s is 0. That is, if $t \in \Sigma$, then we must show that $t = 0$ is formally derivable whenever $t \equiv 0$.

Moreover, it is easy to show that for any term $t \in \Sigma$ there are terms $s_1, \ldots, s_k \in \Sigma$ and polynomials $p_1, \ldots, p_k$ in n variables, with coefficients in $\mathbb{C}$ (also realized as terms in Σ) so that the identity

$$t = p_1 \cdot \exp(s_1) + \cdots + p_k \cdot \exp(s_k)$$

is provable from the permitted axioms. We will prove the theorem by induction on the total number of symbols in the sequence $s_1, \ldots, s_k$, showing that $p_1, \ldots, p_k$ are polynomials, $s_1, \ldots, s_k \in \Sigma$ and $p_1 \exp(s_1) + \cdots + p_k \exp(s_k) \equiv 0$, then $p_1 \cdot \exp(s_1) + \cdots + p_k \cdot \exp(s_k) = 0$ is formally derivable. (Note that we allow s_j to be 0.)

First suppose that $k = 1$: If $p_1 \cdot \exp(s_1) \equiv 0$, then $p_1 \equiv 0$. It is well known that $p_1 = 0$ is provable from the admitted axioms, since p_1 is a polynomial. Hence, $p_1 \exp(s_1) = 0$ also is provable.

From now on assume $k > 1$. Assume $p_1, \ldots, p_k$ are polynomials, $s_1, \ldots, s_k \in \Sigma$ and $p_1 \cdot \exp(s_1) + \cdots + p_k \cdot \exp(s_k) \equiv 0$. For $1 \leq j \leq k$, let π_j be the function on $\mathbb{C}^n$ defined by $\exp(s_j)$. (Choose n so that all variables in each p_j and s_j are included among $x_1, \ldots, x_n$.) Note that we may assume each π_j is nowhere equal to 0 on $\mathbb{C}^n$.

After dividing by π_k we have

$$p_1(\pi_1/\pi_k) + \cdots + p_{k-1}(\pi_{k-1}/\pi_k) \equiv -p_k.$$

Suppose first that we can apply the Hiromi-Ozawa Lemma. In this setting, this means $T(r, \pi_i/\pi_k) \neq O(\log(r))$ for each $1 \leq i \leq k-1$. If so, then there exist constants $c_1, \ldots, c_{k-1}$ (not all 0) so that

$$c_1 p_1(\pi_1/\pi_k) + \cdots + c_{k-1} p_{k-1}(\pi_{k-1}/\pi_k) \equiv 0,$$

which gives us an identity with $k-1$ exponentials after multiplying through by π_k. By the induction hypothesis, the formal identity

$$c_1 p_1 \exp(s_1) + \cdots + c_{k-1} p_{k-1} \exp(s_{k-1}) = 0$$

is derivable in the allowed system. Now we can use this identity to solve for one of the expressions $p_j \cdot \exp(s_j)$ $(1 \leq j \leq k-1)$ and eliminate it from the original expression $p_1 \exp(s_1) + \cdots + p_k \exp(s_k)$. The resulting identity (setting this expression $= 0$) has at most $k-1$ exponentials, so it is derivable. From this one deduces the desired identity

$$p_1 \exp(s_1) + \cdots + p_k \exp(s_k) = 0.$$

On the other hand, it may happen that for some $i(1 \leq i \leq k-1)$, $T(r, \pi_i/\pi_k) = O(\log r)$. Since π_i, π_k are nowhere 0, it follows that $\pi_i \equiv c\pi_k$ for some constant c. [By (C2.3) the same kind of "big-0" estimate holds for π_k/π_i, and hence both π_i/π_k and π_k/π_i are polynomials.] That is, $\exp(s_i) \equiv c \cdot \exp(s_k)$, so that for some constant $d \in \mathbb{C}$, $c = e^d$ and $s_i - s_k \equiv d$. Using the induction hypothesis, we therefore get a formal derivation of $s_i - s_k - d = 0$ and, hence, also of $\exp(s_i) = c \cdot \exp(s_k)$. This allows us to reduce the original identity to one involving only $\exp(s_j)$ for $i \leq j \leq k-1$, which will be derivable by the induction hypothesis. Again this yields a derivation of the identity $p_1 \exp(s_1) + \cdots + p_k \exp(s_k) = 0$ and completes the proof.

Theorem 24.3 has an interesting corollary for trigonometric functions, which we present next. Consider terms in a language with constants for all the complex numbers, variables $x_1, x_2, \ldots$, and function symbols for addition, multiplication, and for sin and cos. Let Σ^* be the set of all these terms.

Corollary 24.4. *If t, s are any two terms in Σ^* and $t \equiv s$, then the identity $t = s$ is provable from the axioms*

$$x + (y + z) = (x + y) + z, \qquad x(yz) = (xy)z,$$
$$x + y = y + x, \qquad\qquad xy = yx,$$
$$x + 0 = x, \qquad\qquad 1 \cdot x = x,$$
$$x(y + z) = xy + xz, \qquad 0 \cdot x = 0,$$

$$\sin(x + y) = \sin(x)\cos(y) + \cos(x)\sin(y),$$
$$\sin(-1 \cdot x) = -1 \cdot \sin(x)$$

together with all axioms giving the facts of addition, multiplication, sin, and cos for constants from $\mathbb{C}$.

Proof. We use the fact that in the context of the complex plane, e^x is interdefinable with sin and cos. Note that since the allowed axioms include the identities $\sin(\pi/2) = 1$ and $\cos(\pi/2) = 0$, we can prove $\cos(x) = \sin(x + \pi/2)$. This in turn allows us to derive the other addition identity,

$$\cos(x + y) = \cos(x)\cos(y) - \sin(x)\sin(y).$$

In Σ^*, let EXP(x) be an abbreviation for the term $\cos(-i \cdot x) + i \cdot \sin(-i \cdot x)$. It is easy to verify that from the allowed identities in Σ^* one can prove the exponential identity

$$\text{EXP}(x + y) = \text{EXP}(x) \cdot \text{EXP}(y)$$

as well as all the numerical facts involving EXP.

Given any term t in Σ^*, we define a term $t^\#$ in Σ by replacing (inductively) each term of the form $\sin(s)$ by

$$-.5i \cdot (\exp(i \cdot s) - \exp(-i \cdot s)),$$

and $\cos(s)$ by

$$.5 \cdot (\exp(i \cdot s) - \exp(-i \cdot s)).$$

If t is any term in Σ we define t^* in Σ^* by replacing (inductively) each term of the form $\exp(s)$ by $\mathrm{EXP}(s)$. Note that if $t \in \Sigma^*$, then the identity $t = (t^\#)^*$ is provable from the axioms allowed in Corollary 24.4.

Now suppose $t, s \in \Sigma^*$ and $t \equiv s$. Then $t^\# = s^\#$, so the identity $t^\# = s^\#$ is provable from the axioms allowed in Theorem 24.3. Hence, $(t^\#)^* = (s^\#)^*$ is provable in the system of Corollary 24.4. It follows that $t = s$ is also provable in that system, completing the proof.

Remark. Suppose $t, s \in \Sigma^*$ and t, s only contain *real* constants. We do not know if there is a proof of the identity $t = s$ in the system of Corollary 24.4 in which *only* real constants appear.

Next we settle positively a conjecture of Schanuel.

Theorem 24.5. *Let $t \in \Sigma$ and suppose the function represented by t is nowhere equal to 0. Then $\log(t)$ is in Σ, in the sense that $t \equiv e^s$ for some $s \in \Sigma$.*

Proof. Let π be the function (on $\mathbb{C}^n$ say) defined by t. There is some holomorphic function G on $\mathbb{C}^n$ so that $\pi \equiv e^G$. We may suppose t is a term of the form $p_1 \exp(s_1) + \cdots + p_k \exp(s_k)$, and we argue by induction on the number of symbols in $s_1, \ldots, s_k$ as in the proof of Theorem 24.3. Clearly we are done if $k = 1$. Assume $k > 1$, and for $1 \leq j \leq k$ let π_j be the function on $\mathbb{C}^n$ defined by $\exp(s_j)$. Then we have $p_1 \pi_1 + \cdots + p_k \pi_k \equiv e^G$ so that $p_1(\pi_1 e^{-G}) + \cdots + p_k(\pi_k e^{-G}) \equiv 1$. We may assume the functions $p_1 \pi_1 e^{-G}, \ldots, p_k \pi_k e^{-G}$ are linearly independent (otherwise, we could replace t by a simpler term to which the induction hypothesis would apply). Hence, the Hiromi-Ozawa Lemma cannot apply. It follows as argued in the proof of Theorem 24.3 that there must exist $1 \leq i < j \leq k$ so that π_i/π_j is identically constant. Again this permits us to reduce the complexity of t and to apply the induction hypothesis. This completes the proof.

We conclude with a related problem.

Problem. *Suppose that f is an entire function for which there exists a term $t \in \Sigma$ such that the function represented by t is equal to f^2. Then must there exist a term $s \in \Sigma$ such that the function represented by s is f?*

Put more simply (but not as correctly), if f is entire and $f^2 \in \Sigma$, must $f \in \Sigma$? Even if one assumes that f^2 and f^3 belong to Σ (and hence $f^n \in \Sigma$ for $n = 2, 3, 4, 5, \ldots$), does it follow that f (assumed to be *entire*) lies in Σ?

References

1. Apostol, T., *Mathematical Analysis*, second edition, Addison-Wesley, Reading, MA, 1974.
2. Beck, W., *Efficient quotient representations quotient representations of meromorphic functions in the disk*, Ph.D. Thesis, University of Illinois, Urbana, IL, 1970.
3. Becker, J., Henson, C. W., and Rubel, L. A., *Annals of Math.* First-order conformal invariants, **112** (1980), 123–178.
4. Berenstein, C., Chang, D.-C., and Li, B. Q., *Complex Variables* A note on Wronskians and linear dependence of entire functions in $\mathbb{C}^n$, **24** (1993), 131–144.
5. Boas, *Entire Functions*, Academic Press, New York, 1954.
6. Bucholtz, J. D. and Shaw, J. K., *Trans. AMS* Zeros of partial sums and remainders of power series, **166** (1972), 269–184.
7. Buck, R. C., *Duke Math J.* Integral valued entire functions, **15** (1948), 879–891.
8. Dienes, *The Taylor Series*, Dover, 1957.
9. Edrei, A., and Fuchs, W. H. J, *Trans. Amer. Math. Soc.* Meromorphic functions with several deficient values, **93** (1959), 292–328.
10. Gauthier, P. M., and Hengartner, W., *Annals of Math.* The value distribution of most functions of one or several complex variables, **96**(2) (1972), 31–52.
11. Gurevic, R. H., *Trans. Amer. Math. Soc.* Detecting Algebraic (In)Dependence of Explicitly Presented Functions (Some Applications of Nevanlinna Theory to Mathematical Logic), **336** (1) (1993), 1–67.
12. Hayman, W. K., *Meromorphic functions*, Oxford, at the Clarendon Press, 1964 (Oxford Mathematical Monographs).
13. Henson, C. W., and Rubel, L. A., *Trans. Amer. Math. Soc.* Some applications of Nevanlinna theory to mathematical logic: identities of

exponential functions, **282**(1) (1984), 1–32 (and Correction 294 (1) (1986), 381).

14. Hille, E., *Ordinary Differential Equations in the Complex Domain*, John Wiley and Sons, New York, 1976.

15. Hiromi, G., and Ozawa, M., *Kōdai Math. Sem. Report,* On the existence of analytic mappings between two ultrahyperelliptic surfaces, **17** (1965), 281–306.

16. Kujala, R. O., *Bull. Amer. Math. Soc.* Functions of finite λ-type in several complex variables, **75** (1969), 104–107.

17. Kujala, R. O., *Trans. Amer. Math. Soc.* Functions of finite λ-type in several complex variables, **161** (1971), 327–258.

18. Laine, I., *Nevanlinna Theory and Complex Differential Equations*, W. de Gruyter, Berlin, 1993.

19. Lang, S., and Cherry, W., *Topics in Nevanlinna Theory*, Lecture Notes in Math. 1443, Springer-Verlag, New York, 1980.

20. Lindelöf, E., *Ann. Scient. Éc. Norm. Sup.* Fonctions entières d'ordre entier, **41** (1905), 369–395.

21. Mahler, K., *Lectures on Transcendental Numbers*, Lectures Notes in Math. 356, Springer-Verlag, Berlin, 1976.

22. Malliavin, P., and Rubel, L. A., *Bull. Soc. Math. France* On small entire functions of exponential type with given zeros, **89** (1961), 175–206.

23. Markuševič, A. I., *Entire Functions*, American Elsevier Publishing Company, New York, 1966.

24. Miles, J., *J. Analyse Math.* Quotient Representations of Meromorphic Functions, **25** (1972), 371-388.

25. Miles, J., *Bull. Amer. Math. Soc.* Representing a meromorphic function as the quotient of two entire functions of small characteristic, **71** (1970), 1308–1309.

26. Nevanlinna, R., *Le Théorème de Picard-Borel lemma et la Théorie des Fonctions Méromorphes*, second edition, Chelsea Publishing Company, New York, 1974.

27. Nienhuys, J. W., and Thieman, J. G. F., *Proc. Dutch Academy of Science, Ser. A.* On the existence of entire functions mapping countable dense sets on each other, **79** (1976), 331-334.

28. Okada, Y., *Science Rep. of the Tôhoku Imperial University* Note on Power Series, **11** (1922), 43–50.

29. Painlevé, P., *Leçons sur la Théorie Analytique des Equations Différentielles*, Profesées à Stockholm, Hermann, Paris, 1897.

30. Palais, R. S., *Amer. J. Math.* Some analogues of Hartogs' theorem in an algebraic setting, **10** (1978), 387–405.

31. Pólya, G., *Math. Zeitschrift* Untersuchungen über Lücken und Singularitäten von Potenzreihen, **29** (1929), 549–640.

32. Reinhart, G., *Schanuel Functions and Algebraic Differential Equations*, Ph.D. Thesis, University of Illinois, Urbana, IL, 1993.

33. Robinson, R., *Trans. Amer. Math. Soc.* Undecidable rings, **70** (1951), 137–159.

34. Rogers, H., *Theory of Recursive Functions and Effective Computability*, McGraw-Hill, New York, 1967.

35. Rubei, L. A., *Duke Math. J.*, A Fourier series method for entire functions, **30** (1963), 437–442.

36. Rubel, L. A., and Taylor, B. A., *Bull. Soc. Math. France* A Fourier series method for meromorphic and entire functions, **96** (1968), 53–96.

37. Rubel, L. A., and Yang, C. C., *Values shared by an entire function and its derivative*, Lecture Notes in Mathematics, Springer-Verlag, Berlin, 1977, pp. 101–103; in Complex Analysis Conference in Lexington, Kentucky, 1976.

38. Rudin, W., *Real and Complex Analysis*, McGraw-Hill, New York, 1974.

39. Schonfield, J. R., *Mathematical Logic*, Addison-Wesley, Reading, MA, 1967.

40. Sodin, M., *Ad. Soviet Math.* Value Distribution of Sequences of Rational Functions, **11** (1992).

41. Stoll, W., *Proc. Sympos. Pure Math.* About entire and meromorphic functions of exponential type, **11** (1968), Amer. Math. Soc., Providence, RI, 392–430.

42. Stoll, W., *Internat. J. Math. Sci.* Value distribution and the lemma of the logarithmic derivative in polydisks, (4) (1983), 617–669.

43. Taylor, B. A., *Duality and entire functions*, Thesis, University of Illinois, 1965.

44. Taylor, B. A., *Proc. Sympos. Pure Math.* The fields of quotients of some rings of entire functions, **11** (1968), Amer. Math. Soc., Providence, RI, 468–474.

45. Tsuji, M., *Japanese J. Math.* On the distribution of zero points of sections of a power series, **1** (1924), 109–140.

46. Tsuji, M., *Japanese J. Math.* On the distribution of zero points of sections of a power series III, **1** (1926), 49–52.

47. van den Dries, L., *Pacific J. Math.* Exponential rings, exponential polynomials, and exponential functions, **113**(1) (1984), 51–66.

48. Vitter, A., *Duke Math. J.* The lemma of the logarithmic derivative in several complex variables, **44** (1977), 89–104.

49. Wilkie, A., *On exponentiation—a solution to Tarski's High School Algebra Problem*, preprint, ca. 1982, but never published.

50. Wilkie, A., private communication.

51. Zygmund, A., *Trigonometrical Series*, 2nd. Ed., Cambridge, 1988.

Index

Universitext   *(continued)*

Sagan: Space-Filling Curves
Samelson: Notes on Lie Algebras
Schiff: Normal Families
Shapiro: Composition Operators and Classical Function Theory
Smith: Power Series From a Computational Point of View
Smoryński: Self-Reference and Modal Logic
Stillwell: Geometry of Surfaces
Stroock: An Introduction to the Theory of Large Deviations
Sunder: An Invitation to von Neumann Algebras
Tondeur: Foliations on Riemannian Manifolds